Sam Illingworth

Science Communication Through Poetry

 Springer

Sam Illingworth ⓘ
Edinburgh Napier University
Edinburgh, UK

ISBN 978-3-030-96828-1   ISBN 978-3-030-96829-8   (eBook)
https://doi.org/10.1007/978-3-030-96829-8

This Springer imprint is published by the registered company Springer Nature Switzerland AG
The registered company address is: Gewerbestrasse 11, 6330 Cham, Switzerland

Science Communication Through Poetry

"This volume blurs the line between Art and Science by showcasing how the understanding and presentation of science using poetry can be artistic and systematic. Students, researchers, practitioners, and poets will find much to admire in this accessible text. Illingworth introduces the reader to a way of reading, analysing, and writing (RAW) science poetically with exercises that both experienced and novice poets will find useful. The step-by-step suggestions for workshops exist at the intersection between scientific facts and poetry's rhythms showing us how to be poetic scientists and scientific poets. This text makes science and poetry clear, exciting, and forges a fruitful path for science communication."

—Sandra L. Faulkner, *Professor of Media and Communication and author of Poetic Inquiry: Craft, Method and Practice*

"*Science Communication through Poetry* by Sam Illingworth is the science communication book you didn't know you needed—but you do. Clearly and eloquently written, the book takes the reader through a journey of how and why poems, whether lines of iambic pentameter gleaned from Shakespeare or a lewd ditty on a pub toilet door, can become useful science communication tools. Thoughtful about the potential exclusionary pitfalls of combining poetry and science, and joyfully ambivalent about poems as 'high literature', this book is a great addition to your science communication toolkit."

—Emily Dawson, *Associate Professor in Science & Technology Studies and author of Equity, Exclusion & Everyday Science Learning*

"*Science communication through poetry* is a brilliantly written book that explores how poetry can communicate science effectively to new audiences, spark discussion, and diversity science. I finished the book feeling inspired that not only that poetry is a powerful way of communicating science, but that poetry—like science—should be for everyone. The book is a real eye-opener about the value of science-poetry collaborations in research. Science and poetry might seem like polar opposites but having read the book I leave with a sense that they are like yin and yang: two seemingly opposite, yet complementary forces, that somehow fit strikingly well together. This book provides the reader with a practical step-by-step guide to understanding how poetry can be used to communicate science without feeling that they've wandered off out of their comfort zone. Having never written a poem since primary school, I felt empowered to go off and write a poem about my own research, and honestly it felt like the most creative thing I've done! And it gave me a buzz and reminded me *why* I do my research. This book is a must-read for anyone who's a scientist, science communicator, a poet, or interested in collaborations in science."

—Gary Kerr, *Associate Professor in Festival and Event Management*

"Disruption comes in a variety guises. Whereas often associated with technological breakthroughs, it may also start as a radically new way of seeing and doing things. Dr. Sam Illingworth is the contributor of such a disruptive practice when he establishes poetry as a powerful language for the communication of science and technology.
In his book *Science Communication through Poetry*, Dr. Sam Illingworth—physicist, science communicator, and poet—explores how the medium of poetry may be used to communicate science effectively—both to and with non-scientific audiences. It is not about replacing one ostracising, excluding, and alienating vernacular with another; instead, Dr. Illingworth demonstrates how poetry can convey not only wonder and emotional engagement but also clarity and insight.
Rather than being a collection of essays on literary theory, this is a very practical handbook. Dr. Illingworth is addressing the poetry- and science-interested reader directly, giving solid hands-on advice on how to find inspiration, write science poetry, and share these poems. Using a number of interesting and captivating examples, he is teaching the reader how to analyse poems, write poems, and perform poems. In his innovative method for poetic transcription, he shows how to formulate

research questions, select data, code and categorise in order to create a piece of science poetry. The final chapters contain some case studies as well as A Manifesto for Poetic Collaborations.

With *Science Communication through Poetry*, Dr. Sam Illingworth opens the door to a new arena of communication and collaboration—a meeting place for professionals who traditionally have been exploring the world along different paths but may have more in common than they have formerly realised."

—Olle Bergman, *Freelance writer (STEM), communication trainer (Academia), and author (popular history), based in Sweden. Member of the Board of the Swedish Authors' Fund.*

For Rebecca & Cora, without whom no poem is complete.

Acknowledgments

This book would not have been possible without the help of *many, many* people.

Thank you to my friends and family for your constant encouragement and for nodding sagely when I try to explain what it is that I do for a living.

Thank you to all of the wonderful colleagues who have inspired me, and from whose work I have drawn substantially. I hope that I have done so in a manner that is both valuable and respectful.

Thank you to the poetry and spoken word communities for welcoming me with open arms, and in particular to my long-time collaborator and friend Dan Simpson for an interdisciplinary association that is worth celebrating.

Thank you to Maria Loroño-Leturiondo for creating all of the wonderful figures that feature throughout this book, and which help to give life to my arguments. Thank you also to Maria for the amazing cover design which perfectly encapsulates the central thesis of the book.

Thank to my handling editor Gonzalo Cordova for having faith in me to deliver this manuscript when I originally pitched the idea, to the wonderful team at Springer Nature for their editorial guidance and expertise, and to Martin Hargreaves for doing such a fantastic job indexing this book.

Finally, I would like to extend a special thanks to the following people, all of whom provided invaluable feedback on earlier versions of the manuscript: Olle Bergman, Emily Dawson, Matthew Dunstan, Leila Howl, Jennifer Roche, and Emma Weitkamp.

Any wisdom or inspiration that you might find in this book are undoubtably grounded in the work of these and others, while any mistakes or missteps remain mine and mine alone.

Contents

About the Author

Dr. Sam Illingworth is an Associate Professor in Academic Practice at Edinburgh Napier University in the UK. His research and practice are concerned with using poetry as a tool to engender dialogue between different audiences, and in particular to give voice to marginalised or underserved communities. You can find out more about his work by visiting his website www.samillingworth.com and connect with him on Twitter @samillingworth.

Chapter 1
Introduction

1.1 Introduction

Science is an essential tool for the development of a healthy, informed, sustainable, and balanced society. It is also a tool that has an unfortunate history of being used to ostracise, alienate, and exclude; a weapon of choice for those who wish to maintain the status quo of a western, patriarchal hegemony of thought [1–3]. Science communication provides a way through which to open up science to a more diverse audience, helping us to move beyond the wrongful exclusion of others from scientific discourse. Not simply because science should be seen as a right rather than a privilege, but because truly diversifying science is the only way to facilitate original solutions to the global interdisciplinary problems that science itself is committed to solving.

I have written elsewhere [4] about the need for inward-facing and outward-facing science communication. That there is a need for science to be communicated to other scientists via peer-reviewed research articles, conference presentations, and academic monographs (inward-facing). How there is also a need for science to be communicated to non-scientists via policy documents, science festivals, and collaborative workshops (outward-facing). This outward-facing side of science communication exists on a spectrum, with dissemination (unidirectional from scientists to non-scientists) at one end, and dialogue (multidirectional between scientists and non-scientists) at the other.

I have deliberately chosen the term dissemination to describe the one-way transfer of information from scientists to non-scientists, rather than 'deficit model', in which scientists are perceived to fill the knowledge 'deficit' of non-scientists. This term holds negative implications about both the intellectual capacity of an audience and the potential motivations of scientists [5], and as such I instead choose to focus on dissemination as a necessary mode of communication which provides reliable information in an accessible way, and which is itself often an essential prerequisite for dialogue [6].

S. Illingworth, *Science Communication Through Poetry*,
https://doi.org/10.1007/978-3-030-96829-8_1

This book aims to explore how we might communicate science effectively both to and with non-scientific audiences across this spectrum of science communication, from dissemination to dialogue, via the medium of poetry.

At this point I imagine that you have two burning questions that you would like me to address. Firstly: is the best way to diversify science and stop it from ostracising, excluding, and alienating to use a medium which is perceived by many to be similarly ostracising, excluding, and alienating? And secondly: who exactly is Sam Illingworth to be telling me how to use poetry to help communicate science?

Well, to answer the second of these questions first: who exactly am I? A scientist? A poet? Both? Neither? In all honesty, these are questions that I have struggled with throughout my academic career, in part because I don't think that I am necessarily the person to make such judgements, and in part because I dislike labels, and find them to be constricting. If I write a poem and keep it in a journal where only I can see it does that make me any less of a poet than if I win the Pulitzer Prize for Poetry? Similarly, if I spend a long summer weekend making detailed observations of all the insects in my garden does that make me any less of a scientist than a recipient of the Nobel Prize in Physiology or Medicine?

So rather than tell you about what I am and what I am not, let me instead tell you a little bit about my academic journey, and how I came to be in a position where I am writing this book on how to communicate science via the medium of poetry.

After a Master of Physics in Physics with Space Science & Technology I embarked upon a PhD in Atmospheric Sciences (both at the University of Leicester in the UK), investigating how satellites could be used to make detailed observations of greenhouse gases at the Earth's surface [see e.g. 7–9]. During my PhD I was involved in a large amount of outreach and public engagement [10] mainly driven by my interest in theatre (I was also the President of the University of Leicester's Theatre Society), and after graduating I was fortunate to receive a scholarship from the Daiwa Anglo-Japanese Foundation to investigate the relationship between science and theatre in Japan. During this scholarship I continued to develop my expertise in science communication and realised that I was perhaps best placed to do so from within academia, and so I returned to the UK to take up a postdoctoral research position, followed by a Lectureship in Science Communication. Shortly after becoming a Lecturer I started writing poetry as a way to try and bring scientific research to new audiences via a blog (see Chap. 2), which I continue to update on a weekly basis. Since then my poetry has appeared in several poetry journals and anthologies, I have performed in spoken word events and venues across the world (from Google's HQ in Silicon Valley to the Green Man Music Festival in Wales), published my research on the relationship between science and poetry in high-impact, peer-reviewed journals [see e.g. 11–13] and edited several collections of poetry and spoken word. I am now an Associate Professor at Edinburgh Napier University, where my research involves using poetry to develop dialogue between different communities. I will return to some of my current and previous research throughout this book, as both exemplars and as learning opportunities from my many mistakes. For now though, I hope that this brief biography serves to illustrate that I am someone for whom poetry has proven to be an effective medium through which to communicate science.

So, that's me. Now to return to the other question: why use poetry to help communicate and diversify science when poetry itself has also been associated with a lack of open communication and diversity?

One of the most common experiences that I encounter when running poetry workshops is when the participants say: 'poetry isn't for me.' When asked to expand on this statement, they usually reveal that this attitude stems from encountering a specific type of poetry while they were at school, which they did not enjoy, and the misled notion that all other poetry must be similar. At this point I like to draw on an analogy. I enjoy many different varieties and genres of music; however, I am really not a fan of Finnish death metal (sorry Finnish death metal fans). I imagine that if I had only ever heard Finnish death metal, I would likely be under the impression that I did not like music.

It is the same with poetry. For many people, they often feel alienated, or excluded, or just plain bored by the poetry that they encounter in their early formative years (I know that this is not true for everyone), and so part of the challenge is in working with these audiences to identify poetry that they do like, and which they feel speaks both to and for them. Doing so helps to reinforce the importance of their own knowledge, expertise, and lived experiences, which in turn helps to ensure agency during the development of subsequent dialogues. By listening to the needs of an audience and working with them to read, analyse, write, share, and perform poetry that they enjoy, it therefore becomes possible to use this as medium through which to open-up and explore science in an engaging, accessible, and stimulating manner.

Another question that I often get asked is: 'Why poetry? Couldn't you just use visual arts, or comedy, or music, or dance to communicate science instead?' Well yes, you could. And there are many great examples of people who do just that [14–17]. However, aside from the fact that I lack any real expertise in any of these disciplines (and am thus ill-suited to write a book extolling the specifics of their various science communication virtues), I believe that poetry offers a unique way to help critique, interrogate, and diversify science. As Salman Rushdie tells us in *The Satanic Verses* [18, p. 97]:

'A poet's work [is] to name the unnameable, to point at frauds, to take sides, start arguments, shape the world and stop it from going to sleep.'

Poetry has the capacity to hold a mirror to science, capturing its reflection warts and all. It is a powerful tool through which to explore the times when science uses the wrong words and ideals, and what we can do to change that. Poetry is also an extremely flexible, adaptable, and accessible medium for communication, providing that it is introduced and framed as such. Throughout this book I will present a large variety of ways in which poetry can be used to give voice to different audiences, the majority of which can be done with nothing more than a pen, a paper, and an open mind.

Individually, science and poetry both present a set of complementary methods that together can help us to better comprehend the world and our place in it, while also diversifying knowledge and understanding. In writing this book I hope to lay out a roadmap for how others might use poetry across the spectrum of science communication, from dissemination to dialogue, with a variety of different audiences.

1.2   How to Use This Book

Following the Introduction, this book is split into seven chapters, each of which provides guidance for different ways in which poetry can be used to communicate science.

Chapter 2—**RAW Poetry** contains an introduction on how to read, analyse, and write science poetry.

Chapter 3—**The Poetry of Science** is concerned with how to turn scientific research into poetry, why you might do so, and how to communicate your work with different audiences across multiple platforms.

Chapter 4—**Poetic Content Analysis** presents a new research method and how to utilise this approach to conduct science communication research.

Chapter 5—**Poetic Transcription** is focussed on another research method involving science and poetry, including worked examples for how to use this in practice.

Chapter 6—**Poetry Workshops** introduces poetry workshops and how they can be used as an effective tool to develop dialogue between scientists and non-scientists.

Chapter 7—**Poetic Collaborations** considers how to develop meaningful collaborations between poets and scientists, drawing on two case studies to consider the roles of the participants in such collaborations.

Finally, Chap. 8 presents **Conclusions** for the work that has been discussed thus far, including potential future directions for science communication research and practice.

Following on from the discussion in Sect. 1.1, Chaps. 2–5 are concerned with using poetry for the largely one-way dissemination of science, with Chaps. 6 and 7 focussed on the use of poetry to engage scientists and non-scientists in multidirectional dialogues. However, the information in Chaps. 2–5 could also be used to help create initiatives that prompt dialogue; likewise for Chaps. 6–7 and the development of strategies for the effective dissemination of science.

As well as containing specific advice and guidance for how to use poetry to communicate science, each of the chapters in this book also contains a number of exercises for you to reflect on what you have learnt and to put into practice what is discussed. There are also sections detailing further study and additional readings that are suggested to help improve your knowledge, understanding, and familiarity with poetry.

In addition to the chapter outlines presented above, I now present an overview for the order in which you might read these chapters, depending with which of the following four identities you most readily associate yourself.

Scientist

If you are a research scientist who is interested in how you might use poetry to develop outreach and public engagement activities for your research, then I would recommend starting with Chaps. 2 and 3. If you think that you would like to develop some of these ideas further and start to run your own poetry workshops then you should read Chap. 6. You might also find Chap. 7 useful for an overview of how to participate in (and develop) effective interdisciplinary collaborations. Having read these chapters and completed their exercises, you will be in a strong position to start using poetry as a way through which to communicate and interrogate your own scientific research. At this stage you could read Chaps. 4 and 5 for an introduction to how poetry can be used as a research method; if you are interested in conducting science communication research, then these chapters provide an outline for how you could adopt a similar approach.

Science Communication Practitioner

If you identify as someone who primarily carries out science communication activities (here I adopt 'science communication' as an umbrella term for outreach, public engagement, widening participation, and knowledge exchange [19]), then you should also start with Chaps. 2 and 3 for a broad introduction to using poetry as a form of dissemination, before moving onto Chap. 6 for an analysis of poetry as a medium for effective dialogue. You might also find that Chap. 7 provides useful information for establishing equitable interdisciplinary partnerships. Chapters 4 and 5 are probably of less use to you than the other chapters but could provide you with inspiration for alternative forms of enquiry into your own practice.

Science Communication Researcher

If you are someone who conducts research in the field of science communication, then the chapters should be read in numerical order. Chapters 2 and 3 introduce using poetry as a method of communicating science to different audiences, forming the foundations for Chaps. 4 and 5 in which I present two new research methods for using poetry in science communication studies. Chapters 6 and 7 then build on the work presented in the previous chapters and describe how poetry can be used as an effectual medium for both dialogue and collaborations between poets and scientists.

Poet

As a poet you will likely find the information in Chaps. 2, , 3, , 6, and 7 to be the most interesting. Depending on your experience, some of the information in Chaps. 2 and 3 will likely be at quite a basic level, but the framing of this information within the context of both science and science communication means that it will still be of value. Similarly, depending on your current (or future) practice, Chaps. 4 and 5 could present alternative methods through which to interrogate and contextualise both your own work and that of other poets.

The above four identities are of course neither mutually exclusive, nor all encompassing. For example, you might be a scientist who also has a long-standing appreciation of poetry, in which case you would likely find some of the information presented in Chaps. 2 and 3 to be familiar. However, given that the majority of this book's readers are likely to come from one of these four audiences I hope that this provides a useful indicator for how to get the most from its contents.

Ultimately this book is meant as a guide to help readers consider how they might use poetry as a way to disseminate science and open up dialogues about science with different audiences. Each of the chapters is based around learnings from my own work and practice, and I hope that it serves as a useful tool for others wanting to communicate science in this way.

Exercise 1.1 Reflect on your exclusion

One of the reasons why science and poetry are seen by many people to be exclusive and 'not for them' is because they have encountered situations in which they have been made to feel excluded. Think of a time when you have been made to feel excluded from something (it doesn't have to be science or poetry) and reflect on how this made you feel at the time. Was your exclusion deliberate? What could have been done differently? Has this had a lasting impact on your engagement with the thing from which you were being excluded?

How might you use your own experiences of being excluded to help ensure that a non-scientific audience can be made to feel included and empowered by science? Likewise, are there specific steps that you can take, based on your own exclusion(s) to make sure that this audience are encouraged and enabled through your use of poetry?

1.3 Summary

This chapter has laid out the basic premise for this book, i.e. that poetry is an effective way to communicate science to a variety of different audiences. It has also introduced the spectrum through which science is outwardly communicated, ranging from the dissemination of information (one-way) to the co-creation of knowledge (multi-way).

By reading the examples presented in this book, completing the exercises, and participating in the opportunities for further study you will have a better understanding of how you can use poetry to communicate science with different audiences, and why it is such a powerful tool to help amplify and diversify science.

1.4 Suggested Reading

For those interested in a more general introduction to science communication as both an academic discipline and practice, I would recommend the wonderful *Science communication: a practical guide for scientists* [20] by Laura Bowater and Kay Yeoman. Those seeking a more in-depth investigation into the history of science communication and an overview of frontier research, including inclusive perspectives on science communication in cultural contexts, should search out the *Routledge Handbook of Public Communication of Science and Technology* [21].

For an excellent introduction into the ways in which the relationships between science and poetry have evolved, I. A. Richards' *Poetries and Sciences* [22] is a fine place to start. Similarly, *The poetry of music and science* [23] by Tom McLeish provides an in-depth consideration of scientific and artistic creativity, while Mary Midgley's *Science and poetry* [24] is a brilliant philosophical text that challenges the concept that science rather than poetry has a 'right' to explaining how the universe operates.

There are also several resources that bring together scientists and poets to discuss the ways in which their disciplines overlap, including *The Measured Word: On poetry and science* [25], an edited collection of essays that explores the relationships between poetic and scientific imaginations, and the *Science Meets Poetry* series [see e.g. 26], which collect the proceedings from the *Science meets Poetry* symposia at the bi-annual European Science Open Forum conference. Finally, *The poetry of Victorian scientists: style, science and nonsense* by Daniel Brown [27] presents an overview of the life of several Victorian scientists who wrote and were influenced by poetry, while my own book *A sonnet to science* [28] provides an insight into six accomplished scientists and the impact that poetry had on their lives and research.

1.5 Further Study

The further study sections at the end of each chapter in this book are an opportunity for you to reflect on what you have learnt and to start to practise some of these learnings and reflections.

The further study section in this Chapter is designed to get you thinking about some of your own experiences of encountering both science and poetry and any potential overlaps that might exist in your reactions to these experiences.

1. **Your favourite science.** What is your favourite piece of scientific research? This could be a fundamental discovery that underpins a specific discipline or field (e.g. genome editing, general relativity, continental drift) or a recent journal article that you read about a particularly engaging project. What is it about this piece of research that speaks to you? Why do you find it to be so inspirational? Make a list of all of the reasons why you have picked this study as your favourite piece of scientific research.

2. **Your favourite poem.** What is your favourite poem? This might be a poem that you learnt in your childhood, or one that you read recently in an online poetry journal. It might be from a multi-award-winning collection, or from an inconspicuous blog on one of the far corners of the Internet. What is it about this poem that speaks to you? Why do you find it to be so inspirational? Make a list of all of the reasons why you have picked this as your favourite poem.

3. **Consider the similarities.** Look at the two lists that you have made and identify any similarities that exist between the two. Are there any overlaps between what inspires you with regards to both science and poetry? What is it about your own expertise and lived experiences that have resulted in these lists? Was one of these lists harder to write than the other one? Why? Taking the time to reflect on the science and poetry that has made a lasting impression in your own life, is a great way to consider how you might use poetry as a way to help others to become inspired and enabled by science.

References

1. Cobern WW, Loving CC (2001) Defining "science" in a multicultural world: Implications for science education. Sci Educ 85(1):50–67. https://doi.org/10.1002/1098-237X(200101)85:1%3c50::AID-SCE5%3e3.0.CO;2-G
2. Saini A (2017) Inferior: the true power of women and the science that shows it. 4th Estate, London
3. Saini A (2019) Superior: the return of race science. 4th Estate, London
4. Illingworth S, Allen G (2020) Effective science communication, 2nd edn. Institute of Physics Publishing, Bristol
5. Simis MJ, Madden H, Cacciatore MA et al (2016) The lure of rationality: Why does the deficit model persist in science communication? Public Underst Sci 25(4):400–414. https://doi.org/10.1177/0963662516629749
6. Trench B (2008) Towards an analytical framework of science communication models. In: Cheng D, Claessens M, Gascoigne T, Metcalfe J, Schiele B, Shi S (eds) Communicating science in social contexts. Springer, Dordrecht
7. Illingworth S, Remedios JJ, Boesch H et al (2011) A comparison of OEM CO retrievals from the IASI and MOPITT instruments. Atmospheric Measurement Techniques 4(5):775–793. https://doi.org/10.5194/amt-4-775-2011
8. Illingworth S, Remedios JJ, Boesch H et al (2011) ULIRS, an optimal estimation retrieval scheme for carbon monoxide using IASI spectral radiances: sensitivity analysis, error budget and simulations. Atmospheric Measurement Techniques 4(2):269–288. https://doi.org/10.5194/amt-4-269-2011
9. Illingworth S, Remedios JJ, Parker R (2009) Intercomparison of integrated IASI and AATSR calibrated radiances at 11 and 12 μm. Atmos Chem Phys 9(18):6677–6683. https://doi.org/10.5194/acp-9-6677-2009
10. Muller CL, Roberts S, Wilson RC et al (2013) The Blue Marble: a model for primary school STEM outreach. Phys Educ 48(2):176–183. https://doi.org/10.1088/0031-9120/48/2/176
11. Illingworth S (2016) Are scientific abstracts written in poetic verse an effective representation of the underlying research? F1000 Research 5:91. https://doi.org/10.12688/f1000research.7783.3
12. Illingworth S, Jack K (2018) Rhyme and reason-using poetry to talk to underserved audiences about environmental change. Clim Risk Manag 19:120–129. https://doi.org/10.1016/j.crm.2018.01.001

13. Illingworth S (2020) "This bookmark gauges the depths of the human": how poetry can help to personalise climate change. Geoscience Communication 3(1):35–47. https://doi.org/10.5194/gc-3-35-2020

14. D'Addezio G (2020) 10 years with planet Earth: the essence of primary school children's drawings. Geoscience Communication 3(2):443–452. https://doi.org/10.5194/gc-3-443-2020

15. Pinto B, Marçal D, Vaz SG (2015) Communicating through humour: A project of stand-up comedy about science. Public Underst Sci 24(7):776–793. https://doi.org/10.1177/0963662513511175

16. Allgaier J (2013) On the shoulders of YouTube: Science in music videos. Sci Commun 35(2):266–275. https://doi.org/10.1177/1075547012454949

17. Myers N (2012) Dance your PhD: Embodied animations, body experiments, and the affective entanglements of life science research. Body Soc 18(1):151–189. https://doi.org/10.1177/1357034X11430965

18. Rushdie S (1988) The Satanic Verses. Viking, New York

19. Illingworth S, Redfern J, Millington S, Gray S (2015) What's in a Name? Exploring the nomenclature of science communication in the UK. F1000Research 4:409. https://doi.org/10.12688/f1000research.6858.2

20. Bowater L, Yeoman K (2012) Science communication: a practical guide for scientists. John Wiley & Sons Ltd., Chichester

21. Bucchi M, Trench B (eds) (2021) Routledge Handbook of Public Communication of Science and Technology, 3rd edn. Routledge, Abingdon

22. Richards IA (1970) Poetries and Sciences: A Reissue with a Commentary of Science and Poetry. WW Norton and Company, New York

23. McLeish T (2019) The poetry and music of science: comparing creativity in science and art. Oxford University Press, Oxford

24. Midgley M (2013) Science and poetry. Routledge, Abingdon

25. Brown K (ed) (2001) The measured word: On poetry and science. University of Georgia Press, Athens

26. Connerade JP, Illingworth S (eds) (2017) Science meets Poetry 5. CreateSpace, Scotts Valley

27. Brown D (2013) The poetry of Victorian scientists: style, science and nonsense. Cambridge University Press, Cambridge

28. Illingworth S (2019) A sonnet to science: scientists and their poetry. Manchester University Press, Manchester

Chapter 2
RAW Poetry

2.1 Introduction

When starting to consider the ways in which poetry might be used to help communicate science, we first need some poems to work with. This Chapter introduces how to find poems about a range of different scientific topics, going on to present an overview of how to read and analyse science poetry before finishing with an outline for writing your own.

I call this a RAW approach to poetry (Reading, Analysing, Writing), and while it might seem entirely obvious to do so, breaking down the handling of poetry into these three distinct phases helps to provide the space that is necessary to engage with it successfully. I have used this approach in workshop sessions for both scientists and non-scientists alike, and while some of the exercises in this Chapter can easily be adapted for running your own poetry workshops (see Chap. 6), I would also encourage you to work through each of them yourself. Doing so will help to give you a better grounding for the ways in which reading, analysing, and writing poetry can be used to help communicate science.

The purpose of this book is to help demonstrate the powerful role that poetry can play in communicating science; however, developing a personal engagement with poetry also brings with it a number of benefits to you as an individual, in both a personal and a professional capacity. Writing poetry has been shown to have physical and mental benefits, with expressive writing found to improve immune system and lung function, diminish psychological distress, and enhance relationships [1]. Poetry has long been used to aid different mental health needs [2], develop empathy [3], and reconsider our relationship with both natural and built environments [4]. As I will discuss further in Sect. 2.5, poetry is also an incredibly effective way of actively targeting the cognitive incubation period [5], improving your productivity and scientific creativity in the process. In short, poetry has a lot to offer, providing you give it the opportunity to do so.

The purpose of this Chapter is therefore to enable you, the reader, to become more comfortable with finding, reading, analysing, and writing science poetry. Doing so will provide you with some of the manifold benefits that I have just discussed and will also help to develop your confidence and understanding in how to use poetry to communicate science with different audiences.

2.2 Finding Poetry

As I outlined in Chap. 1, one of the biggest difficulties in engaging different audiences with poetry is finding the right poems. In doing so we must first acknowledge who our audiences actually are. What are their needs, their experiences, their attitudes towards both poetry and science? Do they have any specific barriers for engaging with poetry, and are there any unique opportunities that we might work with?

The first step towards understanding our audience is in acknowledging that there is no such thing as a 'general' public [6]. Similarly, while I have used the term 'non-scientific' to differentiate from audiences that identify themselves as scientists, I don't imagine that this is an identity that anyone consciously adopts. Despite my protestations, I am not (nor have I ever been) a Premier League footballer, but to my knowledge I have never introduced myself as 'Sam, the non-Premier League footballer'.

Likewise, non-scientists do not identify themselves as such. Depending on their social setting, people may identify themselves according to their jobs, their hobbies, their religious beliefs, their responsibilities as carers, their dietary requirements, their star signs, or any number of things that they actually *are*. Understanding this and getting to know the likes and dislikes of your intended audience, as well as their lived experiences and any cultural sensitivities, is essential for finding the right poems with which to engage them.

> **Exercise 2.1 Who are you?**
>
> Let's start with you. Try and write a list of all of the different things that you identify as.
>
> I'll go first. I'm a husband, a father, a scientist, a Christian, a poet, and a gamer. I'm also a runner, a games designer, a Leeds United fan, and a very apologetic Brit. These are some of the identities with which I associate myself; only one of which is a scientist.
>
> Understanding the different identities that you associate with yourself, along with your likes and dislikes helps to reinforce the notion that there is no such thing as a general public, but rather that there are many different communities, none of whom primarily identify themselves as non-scientists. It also helps you to think about the kind of poetry that you might want to engage with in the first instance.

In terms of actually finding poems, and science poems in particular, the Internet is the best place to start. Websites such as the *Poetry Foundation* [7] and *Poets.org* [8] have tens of thousands of poems that you can search through according to keyword or topic. *Consilience* [9] (see also Chap. 7), *The Sciku Project* [10], and my blog *The Poetry of Science* [11] also contain several hundred poems that are specifically related to scientific topics ranging from climate change and animal conservation to quantum biology and space exploration.

In terms of print publications, *Magma*, *The Rialto*, *Rattle*, and *The Poetry Review* all offer periodicals that provide a platform for a variety of different poems and poets. Similarly, *Litmus* [12] is a semi-regular publication exploring the interaction between poetry and science. With any of these publications it is worth visiting their website in the first instance to see if the kind of poems that they publish appeal to you.

Collections of poetry that would be welcome on almost any bookshelf include *Staying Alive: poems for unreal times* [13] and *Being Alive* [14], both of which contain poems that reflect on science and the relationship between scientists and society. Similarly, *A Quark for Mister Mark: 101 poems about science* [15] provides exactly what its title suggests, with a selection of science-themed poems from Ancient Greece through to the modern era.

As you have probably noticed, all of these recommendations (and indeed all of the poems that feature in this book) are written in (or translated in to) the English language. This is because one of my many failings in life is that I am unable to understand any language other than English to a proficiency that permits me to enjoy poetry. I am working on this, but in the meantime please be assured that every other language and associated cultures have many different examples of unique and wonderful poetry, and if the limits of your own language extend beyond those of mine, then I encourage you to seek them out and revel in their splendour.

Exercise 2.2: Find some poems

Having identified who you are, find some poems that you think you might enjoy. Visit one of the websites that I mentioned in Sect. 2.2 and find some poems that you can use in the exercises throughout the rest of this Chapter. Begin by searching for poems that relate to either your specific field of scientific research or a research topic that you are interested in (e.g. 'migration patterns of butterflies' or 'ocean acidification'), and then broaden out your search to include any related terms.

Remember, you are not necessarily judging, analysing, or even reading the poetry. You're just collecting poems that you think may be of interest, so that you can use them in our RAW (Reading, Analysing, Writing) approach to poetry, in the proceeding sections.

One final note about finding poems. Don't be snobbish or dismissive. If you've identified some sixteenth-century ballads, then brilliant. But it's equally as brilliant if you've identified a hastily scribbled limerick in the cubicle of a dingy night club. Telling people (yourself included) what poetry they are and are not allowed to engage with is simply another way of reinforcing feelings of alienation and exclusivity. Furthermore, engaging with poetry on a personal level means different things to different people, and even those of us with overlapping interests and tastes may respond differently to the same poem.

2.3 Reading Poetry

Read this poem, 'When the Mangroves Disappeared' in your head:

Deep beneath the dunes
fossilised roots whisper
golden memories
of emerald lagoons.
When sapphire seas
lapped tenderly
at knotted feet,
bathing sunken stems
with the tidal surge
of their brackish embrace.

Dredged up alongside
buried treasures,
dark reflections stir.
Open wounds that sing
of their betrayal –
silted, shifting waters
suffocating with
dry and barren soils.

Until their waves
broke in silence
upon the dead
and burning sands.

Now read the same poem out loud, making sure that you are in an environment in which you feel comfortable to do so.

Do you notice any differences in how the poem sounds, or how it 'feels' to be read? I'm not asking you to analyse the poem at this moment in time, but to simply take the time to reflect on how it sounds when it is read in your head and when it is read out loud. Do you prefer either of these?

'When the Mangroves Disappeared' was written by me, and I will be using several of my poems throughout this book for two reasons. Firstly, having written these poems I know exactly what they are 'supposed to be about', which is extremely useful for when we come to analyse them (see Sect. 2.4). The second reason is one of practicalities; I own the copyright for these poems and reproducing the complete works of other poets in books such as this can be an extremely arduous and costly process.

Now read the following poem 'Fishy Traits' (again by me), both in your head and out loud, noting down any differences or preferences according to your method of delivery:

Fish can't shrug,
fish can't cry,
fish cannot get mad;
fish can't sulk
fish can't frown,
or tell us if they're sad.

But fish can turn,
fish can move,
fish can start and stop;
so can we recognise
each fish from
traits that they adopt?

Angry fish,
happy fish,
fish with spiny backs;
sad fish,
frightened fish,
movements we can track.

Fish that don't stay stationary
like to burst with speed,
while those that have a lengthy stride
travel far indeed.

Watching how these fish all swim
might show us how they feel,
a shoal of individuals
whose temperaments are real.

Exercise 2.3: Read some poems
Pick three of the poems that you identified in Exercise 2.2 and read them to yourself, first in your head and then out loud. You don't need to start thinking about what the poems may or may not mean, or even what it is that you do or do not like about them. Simply read them in your head and out loud and note any differences or preferences in their delivery.

In this initial reading you should not be worried about analysing the poetry in any purposeful way, but rather you should be granting yourself the permission to read these poems and to listen to how they sound when they are read in your head, and when they are read out loud.

2.4 Analysing Poetry

From my experience of running science and poetry initiatives, one of the biggest hurdles in encouraging people to engage with science poems is in their analysis. Many people approach analysing poetry with the pre-conceived notion that it is an extremely complex artform, for which years of study and an encyclopaedic knowledge of the poet and their influences are essential. However, it is exactly this attitude which serves to put many people off from engaging with poetry on their own terms, and the realisation that the analysis of poetry can be straightforward, rewarding, and even fun.

This is not to belittle or undermine the work of those researchers and other experts who invest a great deal of effort and intellect in analysing and contextualising poetry, but rather it is to say that there are many ways in which poetry can be analysed. To illustrate this point, let's look at three quotes from exceptionally talented poets, all of which offer opinions on poetry and the ways in which it can be interpreted:

'Poetry lifts the veil from the hidden beauty of the world, and makes familiar objects be as if they were not familiar.' – *Percy Bysshe Shelley*

'Poems are like dreams: in them you put what you don't know you know.' – *Adrienne Rich*

'Poetry is the journal of a sea animal living on land, wanting to fly in the air.' – *Carl Sandburg*

Now, while I greatly respect the work of these three poets, whenever I read statements like these, I can't help but think that they act to alienate rather than inspire others to engage with poetry. When people are shown examples like this, it is unsurprising that they find the notion of analysing poetry to be entirely unappealing. Instead, I prefer the following quote, from a far less talented poet:

'Poetry is whatever it means to you, and how it relates to *your* needs, *your* experiences, and *your* understanding.' – *Sam Illingworth*

The only person who really knows what a poem was written about is the poet who wrote it, but that doesn't mean that the poem cannot be interpreted in any other number of ways, or that any of these interpretations are any more or less 'correct'. One of the most incredibly powerful aspects of poetry (and why it is so effective as a tool for science communication) is because it enables people to engage with its content in an almost infinite combination of possibilities. When you read a poem, you do so while simultaneously bringing all of your own lived experiences, knowledge, and identity to your interpretation, or analysis, of the poem. Therefore, while you might not know exactly what the poet was thinking or feeling when they wrote a particular poem, it is also true that they are extremely unlikely to know the exact circumstances of your own life that would have led up to your reading of it.

Finding out more about the poet and their circumstances of writing will of course lend further context to your interpretation of the poem, but it shouldn't change the fact that you can still have an opinion on a poem, what it is about, and what it means to you without any of this prior knowledge.

To illustrate this point, let's look once more at the poem 'When the Mangroves Disappeared' (Sect. 2.3). Read the poem again (either out loud, or in your head, or both), and ask yourself these three questions as you do so:

1. What do you think this poem is about?
2. How does the poem make you feel?
3. Do you like the poem?

With regards to Question 1, what makes you think that the poem is about this subject? With regards to Question 2 and Question 3, why did you have this reaction? Is there something that you are bringing to the poem when you read it that makes you feel this way, and which colours your appreciation (or not) of the poem.

I should point out that it is, of course, perfectly ok to dislike any (or all) of my poetry; just as it is perfectly ok to detest those poems that are beloved by millions, and to love those poems that others turn their noses up at. With that in mind, really consider what it is that you dislike (or like) about the poem and take the time to analyse why it causes *you* to react in this way.

I wrote 'When the Mangroves Disappeared' for my weekly blog *The Poetry of Science*, in which I pick a recent piece of scientific research and write a poem to try and introduce the work to new audiences. This particular poem was in response to a recent study [16], which has shown that the mangrove forests on the coasts of Oman disappeared because of climate change.

Mangrove forests, or lagoons, occur worldwide, in the tropics and subtropics mainly between latitudes 25° N and 25° S, and fossil records show that there used to be many mangroves on the coast of Oman. However, these disappeared almost entirely approximately 6,000 years ago. Nowadays, the only mangroves in Oman are those of a particularly robust species and they can only be found in a very few places. In this study, researchers compiled numerous geochemical, sedimentological, and archaeological findings to present a more accurate picture of what led to the large-scale collapse of the Oman mangroves.

Their findings indicate that the collapse of these ecosystems was because of a relatively sudden change in both the local and the global climate, caused by a shifting of the Intertropical Convergence Zone (a band of low pressure around the Earth which generally lies near to the equator).

Having read this brief explainer about the scientific study that inspired me to write this poem, ask yourself those same three questions again:

1. What do you think this poem is about?
2. How does the poem make you feel?
3. Do you like the poem?

Now that you know what I wrote the poem about (the disappearance of mangroves in Oman via historical climate change), why I wrote it (to introduce this research to a new audience and encourage them to find out more about it), and who I wrote it for (followers of my blog), how have your responses to these three questions changed? Does knowing some of the context for the poem change how you feel about it (Question 2) or whether you like it or not (Question 3), and if it does, why do you think that this is the case? With regards to Question 1, do you now think that the poem is about something completely different to what you imagined, and if so, does the context that I provided completely negate your own opinions as to what the poem is about?

Specifically, with the notion of communicating science in mind, if you did imagine the poem to be about something other than the disappearance of mangroves in Oman because of historical climate change, could this poem still be used to spark dialogue around the topic that you had suggested? If so, then how is your interpretation of the poem (as the reader) any more or less correct than mine (as the author)?

Let's repeat this experiment with 'Fishy Traits'. First re-read the poem (Sect. 2.3) and ask yourself those same three questions, remembering to consider exactly what it is about your own lived experiences that has elicited this response.

'Fishy Traits' is another poem from *The Poetry of Science*, again written for a non-specialist audience to try and inspire them to find out more about a recent piece of scientific research; in this instance, that the way a fish swims reveals a lot about its personality [17]. In this study, researchers filmed the movements of 15 three-spined stickleback fish swimming in a tank which contained two, three, or five plastic plants in fixed positions.

Using high-resolution tracking data from video recordings, measurements were made of the directions that the fish turned, how often they turned, and how much they stopped and started moving. In analysing these data, the researchers found each of the fish moved in a unique way, and that these differences were highly repeatable; so much so that the researchers could identify an individual fish just from looking at its movement data. This finding suggests that we might be able to quantify personality differences in wild animals simply by getting fine-scale information on the ways in which they are moving, thereby providing a robust way to analyse individual behaviours in species that are otherwise difficult or impossible to study.

Having read this brief explainer with regards to the science that inspired me to write this poem, ask yourself those three questions again. Has anything changed? Why do you think that this might be the case?

Exercise 2.4: Analyse some poems

Use the three poems that you read in Exercise 2.3 and re-read them with these three questions in mind:

1. What do you think this poem is about?
2. How does the poem make you feel?
3. Do you like the poem?

Make notes of your responses for each of these poems and try to establish what it is about each of the poems that causes such reactions, especially with regards to Question 2 and Question 3.

Now try to see if you can find any information that provides additional context for the poem. Are there any interviews with the poet where they explain their inspirations for this piece? Does the poet's biography reveal some further information about why they might have written the poem or who they wrote it for? Does the timing of the poem correspond to any major historical event or scientific discovery? Does the poem itself contain an epigraph (a quotation from another literary work that is typically placed beneath the title at the beginning of a poem), dedication, or any footnotes that might reveal more about the poet's rationale for writing this piece?

Find as much information as you can about the authors, and then re-read their poems while asking yourself these same three questions. Make notes of your responses, and then compare and contrast these to your original responses. Are they different? Why?

Having done some basic research into the poets, what about the scientific topics that the poems relate to? Are these topics that you understand well? Are there any specific words or phrases in the poem that might relate to scientific jargon or phenomena with which you are not familiar? Take the time to do a little reading around the related scientific topics for each of these poems (for example, if you had picked 'When the Mangroves Disappeared' you might want to find out more information about mangrove ecosystems, historical climate change, and the Intertropical Convergence Zone), and then ask yourself these three questions again. Have your responses changed? Why do you think that this is the case?

Figure 2.1 presents a visual summary of this approach to analysing science poems, with the caveat that in some instances researching the poet and their rationale for writing might not always be possible.

But how will analysing poetry in this way help you to communicate science?

Fig. 2.1 A process for
analysing science poems,
considering your own lived
experiences as well as
information about the poet
and the scientific topic in
question

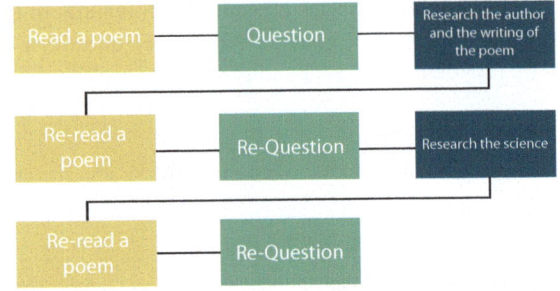

Analysing poems using this process is an effective exercise for engaging different audiences with scientific topics through poetry (see Chap. 6). However, it is also an extremely useful way of helping you to understand your personal attitudes, knowledge, and beliefs in relation to a specific scientific topic.

Let us imagine that you are a scientific researcher who specialises in the acoustics of whale song. Whatever stage in your career you are at, you will have many different experiences in relation to your scientific discipline(s) and your attitudes towards it. Some of these will be scientific (e.g. your undergraduate degree in marine biology, your postgraduate degree in acoustic ecology), some of them will be personal (e.g. your membership of the WWF, hearing whale song for the first time on a nature documentary), and some of them will be a combination of the two (e.g. your first sighting of a blue whale on a scientific expedition, frustrations during the final stages of writing your thesis).

When you come to communicate your scientific research to different audiences, you do so with these embedded experiences. While you may explicitly use some of these experiences and attitudes in your communications, there will be many others that you are less aware of. Reading poems that relate in some way to your scientific topic(s) of interest and analysing them in terms of your own lived experiences (by following the steps shown in Fig. 2.1) helps you to become more aware of these attitudes and biases and to better connect with your subject, including why you chose to engage with it in the first instance. Doing so also helps you to be more open in the way that you communicate your research to others.

If you are a scientific researcher, then aside from the area of research which you may specialise in, you will also have encountered many other areas of science, either in your earlier studies or because you keep abreast of popular scientific topics via a variety of different media (e.g. journals, podcasts, blogs). Alongside these encounters you bring with you other lived experiences. For example, you might be a climate change researcher who is reading an article about a potential cure for Type 2 diabetes, having a parent who lives with this condition.

Similarly, if you are a science communicator who is interested in communicating science more generally, then you will be familiar with many different scientific topics, which in turn will be tempered by your personal and professional experiences. Reading science poems and analysing them as has been outlined in this Section will help you to better understand the ways in which you engage with different scientific topics, which in turn will make you more aware of how (and why) you communicate them with different audiences.

2.5 Writing Poetry

Now that you feel more comfortable with finding, reading, and analysing science poetry, it is time to think about writing some of your own. Chap. 3 will go into more detail with regards to the specific details of writing and sharing science poetry, but this section will provide you with a basic introduction for how to start.

Asking someone to 'write a poem' can be quite an overwhelming request. To me it feels similar to asking someone to 'do some science'. However, introducing some constraints and structure via poetic forms, provides the scaffolding on which creativity can blossom.

A poetic form refers to a type of poem that follows a particular set of rules. For example, the number of lines, the rhyme scheme, the syllable count, etc. There are many different types of poetic form and variations in English-language poetry, and even more in the poems of other cultures and languages. Of course, poems do not have to stick to any of these forms but using them can help to build your confidence as a poet. Similarly, if you want to develop your own structures by breaking existing conventions and rules, it helps to understand what these are in the first instance.

Think of the following scientific analogy: the proponents of quantum mechanics would not have been able to highlight the limitations of classical mechanics without first understanding classical mechanics and how they were derived. It is the same with poetry, you don't have to stick to any form, but understanding some of their rules will give you the confidence and skill to eventually break free and create your own poetic experiments. Furthermore, sometimes picking the right poetic form will really help a scientific topic to be communicated more effectively, as is the case with the first poetic form that I would like to introduce: the nonet.

The origins of the nonet are uncertain, but it is likely that it derives its name from the musical term 'nonet', which refers to a group of nine musicians. It is a nine-line poem, where each line contains a specific, descending number of syllables. The first line contains nine syllables, the second line contains eight, the third line contains seven, and so on, until the last line of the poem, which contains one syllable. The structure for the nonet is shown in Fig. 2.2

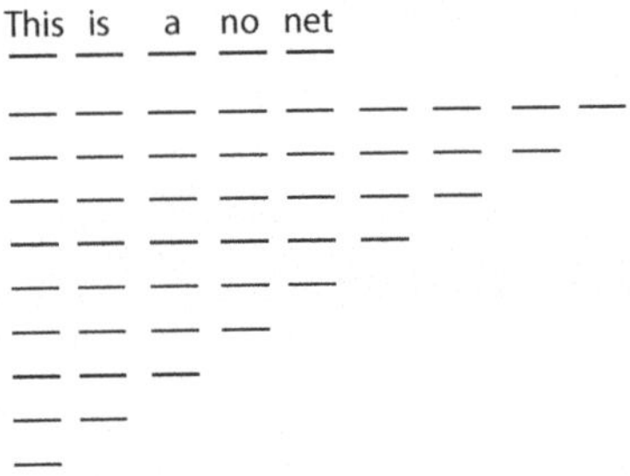

Fig. 2.2 The structure of a nonet

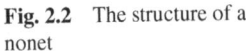

My poem 'Going North for the Winter' provides an example of what a nonet looks like in practice.

I feel the powder beneath my feet;
a sickening howl to the south,
but I know that I am safe
and so I dance with joy.
Until my heart stops,
as I see that
I have run
out of
snow.

This nonet was inspired by research which found that climate change is forcing the habitat of the snowshoe hare (a species of hare found in North America) further north [18], with reduced snow cover meaning that it must shift its habitat to avoid lynx, coyotes, and other predators. As you can see the nonet lends itself well to topics or ideas in which the poet is trying to convey a sense of something diminishing; in this instance the rapidly disappearing habitat of the snowshoe hare, brought about by climate change.

Exercise 2.5: write a nonet
Pick one of the science poems from this chapter and use Fig. 2.2 to write a nonet about a scientific topic that it is addressing. For example, you may choose to use 'Fishy Traits' and to write a nonet about the observable personality differences in wild animals. When you are writing your nonet, try to use the shape of the poem to help tell your story.

The second poetic form that I would like to (re-)introduce you to is the haiku (the plural form for which is also 'haiku'). Haiku originated in thirteenth-century Japan as the opening verse of a 'renga' an oral poem generally a hundred verses long, which was composed collaboratively at social gatherings. The much shorter haiku broke away from renga in the sixteenth century and became popularised in the West after World War II. Many people in the West are now introduced to haiku at school, where they are taught that it is a 3-line poem with a syllable count of 5–7–5 (i.e. 5 syllables on the first line, 7 syllables on the second line, and 5 syllables on the third line). However, this does not really capture the essence of a haiku and how to write one.

A traditional Japanese haiku consists of three lines with a 5–7–5 'kana' count (for a total of 17 kana). The kana is the Japanese language unit, but it is *not* equal to an English-language syllable. Therefore, when writing haiku in English (or any non-Japanese language) getting caught up in syllable counts is unnecessary. Instead, I would recommend making your haiku 17 syllables or less but would also encourage you to concentrate on the other rules that are necessary when composing a haiku.

Haiku should always be about nature (if the haiku you are writing is about human nature, then it is not a haiku, but is a 'senryu' instead). Haiku should also contain a season word. In Japanese there is a strict list of season words (called 'kigo'), but again when writing in English I would advise some flexibility, by picking a word that *you* associate with a season. For example, a seasonal fruit, or flower, or event. You should avoid dual or conflicting seasons wherever possible.

Additionally, a haiku is not written in the past or the future, but rather covers a single event over a short time period; think of it as a live photo (a camera feature available on some smart phones, where a moving image is captured, from both immediately before and after the image was taken) in poetic form.

Finally, haiku should contain juxtaposition, or contrast. In Japanese this can be done by picking from a selection of 'kireji' (or cutting words), but in English this needs to be done in a slightly less subtle fashion, by deliberately introducing juxtaposition in space, time, context, etc. by placing two contrasting objects or images alongside one another.

Figure 2.3 summaries these rules by providing a checklist for writing haiku.

Let us consider one of my own poems 'You all Fall Down', which I offer here as an example for how to construct a haiku:

Balanced on a rock

Orange sneezeweed in your mouth;

You drop in the heat.

Fig. 2.3 A haiku checklist (don't worry too much about the syllable count!)

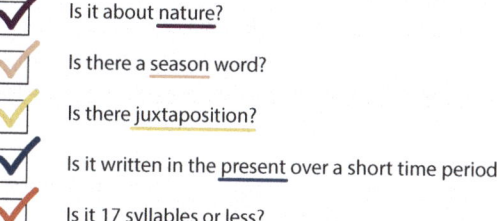

The haiku checklist

☑ Is it about <u>nature</u>?

☑ Is there a <u>season</u> word?

☑ Is there <u>juxtaposition</u>?

☑ Is it written in the <u>present</u> over a short time period?

☑ Is it <u>17 syllables</u> or less?

This haiku was written in response to research which found that climate change has been responsible for the local extinction of the American pika (a small, herbivorous mammal with thick, light brown fur) from California's Sierra Nevada mountains [19]. So how does this haiku compare to the checklist presented in Fig. 2.3?

1. Is it about nature? **Yes.**
2. Is there a season word? **Yes – this species of plant tends to bloom in midsummer.**
3. Is there juxtaposition? **Yes – we are presented with a serene picture of a pika that is daintily perched on a rock before there is a contrast in tone and the animal drops down dead because of heat exhaustion.**
4. Is it written in the present over a short time period? **Yes – the time it takes the pika to fall off the rock.**
5. Is it 17 syllables or less? **Yes. Here I have adopted a 5–7–5 syllable count, but I needn't have done so.**

You can see here that the haiku is a really effective poetic form for capturing the essence of the science that I wanted to communicate. My intentions with this poem were to contrast the cute nature of the American pika with the harsh realities of climate change, and the haiku presents the ideal form for this juxtaposition to be presented to the reader.

Exercise 2.6: write a haiku

Take the nonet that you wrote in Exercise 2.5 and imagine a single scene that occurs during the narrative that you have constructed. Using the checklist presented in Fig. 2.3 as your guide, write a haiku about that scene.

Compare and contrast the two poems. Are they different? How? Which one is more effective at communicating the scientific topic in question? Why?

As I briefly mentioned in Sect. 2.1, another way in which writing science poetry can be beneficial is in actively targeting the incubation period, but how so?

Imagine that you are working on a specific task (scientific or otherwise) and that you reach an impasse, a problem that you are not able to solve. Good practice would have it that one of the ways to overcome this is to step away from your desk (or lab bench or computer or experiment) and put the problem to one side, allowing your subconscious to process the issue and potentially come up with a solution. This is what is known as the incubation period.

However, this is a very passive approach to utilising the incubation period, as by acting in this way you are hoping that your brain will eventually come up with a solution and then tell you about it when you are in the bath/watching TV/going for a run. An alternative approach is to actively target the incubation period, by thinking about the problem in a completely different way; for example, by writing a poem about it. As I will discuss in the proceeding chapters, writing a poem about your own scientific research is an effective way to communicate this work to different audiences, but it also enables you to conceptualise your work from an entirely original perspective, and in so doing presents opportunities for problem solving by actively targeting the incubation period [5]. In other words, poetry is an incredibly powerful tool that all scientists and science communicators should be using more.

2.6 Summary

This Chapter has been designed as a confidence builder for engaging with poetry, providing an introduction for how to find different science poems, as well as guidance for how to read and analyse them in relation to the science that they involve. Some basic scaffolding and structure for creating your own science poems, and the benefits for doing so, have also been discussed. This will be expanded upon in Chap. 3, where we will take a more detailed look at how to write science poems, and then what we might do with them once they are written. By the end of reading this Chapter you should also have a greater appreciation of poetry and the role that it can play in both examining and communicating science.

2.7 Suggested Reading

One of the most accessible and engaging books for opening up poetry to a wider audience is Stephen Fry's *The Ode Less Travelled* [20], which presents a whimsical introduction to writing and appreciating poetry, including many exercises and activities. Similarly, *How to be a Poet* from Jo Bell and Jane Commane [21] features a series of short essays on how to read, write, and enjoy poetry, including additional contributions from a wide variety of internationally-acclaimed poets, all of which are easily digestible and full of practical and actionable advice. As well as the *Staying Alive* [13] and *Being Alive* [14] collections that have already been mentioned in this Chapter, *Poetry by Heart: Poems for Learning and Reciting* [22] contains over 200 English-language poems stretching from the present day all the way back to the Anglo-Saxons. In addition to being a genuine treasure trove of poetic variety, this book also features a detailed description about every poem and their author, which is ideal for experimenting with the approach to poetic analysis that is outlined in Fig. 2.1. Finally, and in addition to those resources already listed in Sect. 2.2, there are many wonderful collections of science poetry from individual poets for you to seek out. Some of my favourites include: *A Responsibility to Awe* by Rebecca Elson [23], *Dung Beetles Navigate by Starlight* by Sarah Watkinson [24], *Scientific Papers* by David Morley [25], and *Elemental Haiku* by Mary Soon Lee [26]. You might also enjoy *A Celestial Crown of Sonnets* [27], a sequence of sonnets that I co-wrote with Stephen Paul Wren, which explore the history of astronomy from antiquity to modernity.

2.8　Further Study

The further study in this Chapter is designed to help you think about how to read, analyse, and write science poems, and to reconsider some of your attitudes and beliefs towards different scientific topics:

1. **Read more poetry**. Consider the three poems that you read in Exercise 2.3, and which you went on to analyse. Was there one of these poems that you particularly enjoyed? Find some more poems written by the same poet and read and analyse these as well. They need not be about science, but if this is the case consider instead the topic that they address, and your own knowledge and attitudes in relation to this.
2. **Challenge yourself**. Think of a scientific topic or subject that you do not enjoy, or in which you have no interest, and find a poem that is related to this. Read the poem and analyse it via the steps outlined in Fig. 2.1, using this to reflect on why you have these attitudes towards this particular topic. Does reading and analysing this poem challenge those views, or lead to any greater understanding as to why you feel this way about that particular scientific topic?
3. **Actively target the incubation period**. The next time that you come across a problem in your work that you are finding difficult to overcome, take a break and write a nonet or a haiku about it. Once you have written your poem, reflect on your problem and see if it leads to any new understanding.

References

1. Lepore SJ, Smyth JM (2002) The writing cure: How expressive writing promotes health and emotional well-being. Am Psychol Assoc, Washington DC
2. Mcardle S, Byrt R (2001) Fiction, poetry and mental health: expressive and therapeutic uses of literature. J Psychiatr Ment Health Nurs 8(6):517–524. https://doi.org/10.1046/j.1351-0126.2001.00428.x
3. Furman R (2005) Using poetry and written exercises to teach empathy. J Poet Ther 18(2):103–110. https://doi.org/10.1080/08893670500140549
4. Elder J (1996) Imagining the earth: Poetry and the vision of nature. University of Georgia Press, Athens
5. Januchowski-Hartley SR, Sopinka N, Merkle BG et al (2018) Poetry as a creative practice to enhance engagement and learning in conservation science. Bioscience 68(11):905–911. https://doi.org/10.1093/biosci/biy105
6. Bultitude K (2011) The why and how of science communication. In: Rosulek P (ed) Science communication. European Commission, Pilsen
7. Poetry Foundation (2021) Poetry Foundation. https://www.poetryfoundation.org. Accessed 10 December 2021
8. Academy of American Poets (2021) Poets.org. https://poets.org/. Accessed 10 December 2021
9. Illingworth S (ed) (2021) Consilience. https://www.consilience-journal.com. Accessed 10 December 2021
10. Holmes AM (2017) Science in 17 syllables. Science 358(6365):966. https://doi.org/10.1126/science.358.6365.966

11. Illingworth S (2021) The Poetry of Science. https://thepoetryofscience.scienceblog.com. Accessed 10 December 2021
12. Lehane D, Cleghorn E (eds) (2014) Litmus: the neurological issue. Litmus Publishing, London
13. Astley N (ed) (2002) Staying alive: real poems for unreal times. Bloodaxe Books Ltd., Highgreen
14. Astley N (ed) (2004) Being alive. Bloodaxe Books Ltd., Highgreen
15. Riordan M, Turney J (eds) (2000) A Quark for Mister Mark: 101 poems about science. Faber and Faber Limited, London
16. Decker V, Falkenroth M, Lindauer S et al (2021) Collapse of Holocene mangrove ecosystems along the coastline of Oman. Quatern Res 100:52–76. https://doi.org/10.1017/qua.2020.96
17. Bailey JD, King AJ, Codling EA et al (2021) "Micropersonality" traits and their implications for behavioral and movement ecology research. Ecol Evol 11(7):3264–3273. https://doi.org/10.1002/ece3.7275
18. Sultaire SM, Pauli JN, Martin KJ et al (2016) Climate change surpasses land-use change in the contracting range boundary of a winter-adapted mammal. Proc R Soc B: Biol Sci 283:20153104. https://doi.org/10.1098/rspb.2015.3104
19. Stewart JA, Wright DH, Heckman KA (2017) Apparent climate-mediated loss and fragmentation of core habitat of the American pika in the Northern Sierra Nevada, California, USA. PLoS ONE 12(8):e0181834. https://doi.org/10.1371/journal.pone.0181834
20. Fry S (2007) The ode less travelled: unlocking the poet within. Arrow Books, London
21. Bell J, Commane J (eds) (2017) How to be a Poet. Nine Arches Press, Rugby
22. Blake J, Dixon M, Motion A et al (eds) (2014) Poetry by heart: Poems for learning and reciting. Penguin Books Ltd., London
23. Elson R (2001) A responsibility to awe. Carcanet Press, Manchester
24. Watkinson S (2015) Dung beetles navigate by starlight. Cinnamon Press, Tanygrisiau
25. Morley D (2002) Scientific papers. Carcanet Press, Manchester
26. Soon Lee M (2019) Elemental haiku. Ten Speed Press, Berkeley
27. Illingworth WSP (2021) A celestial crown of sonnets. Penteract Press, Padstow

Chapter 3
The Poetry of Science

3.1 Introduction

In Chap. 2 I outlined a RAW (Reading, Analysing, Writing) approach to engaging with science poetry, the purpose of which was to help you feel more comfortable in working with poetry and to begin to think about how you might use it to communicate science with different audiences. In this Chapter I will develop some of these ideas further, specifically focussing on the writing of science poetry, before going on to discuss how you might share these poems, and what the benefit might be for doing so.

One of my underlying objectives for this book is to provide a safe yet effective tool for science communication, particularly for those looking to engage with the poetic medium for the first time. As such, by introducing you to a series of techniques, methodologies, and strategies for using poetry to communicate science I hope to enable you to do so, but I also hope that I can make poetry itself more accessible and that this book can give you the confidence, experience, and permission to engage with poetry more frequently.

What follows in this Chapter is thus a framework for helping you to identify a scientific topic, write a poem about it, and share that poem with an audience. This framework is informed by the work of others, but it has also been developed based on my own personal experiences of trying, failing, and succeeding to create poetry that communicates a range of scientific topics. Your own journey to creating and sharing science poetry will no doubt be very different from my own and from those of your colleagues and contemporaries, but by following the guidance in this chapter, completing the exercises, and making the time to do so, you will succeed in creating poetry that both communicates science, and which reaffirms the need to do so.

© The Author(s), under exclusive license to Springer Nature Switzerland AG 2022
S. Illingworth, *Science Communication Through Poetry*,
https://doi.org/10.1007/978-3-030-96829-8_3

3.2   What is a Poem

What is a poem? How does it differ from prose? And how do you know if what you have written is even classified as poetry?

As discussed in Chap. 2, I find some definitions of what poetry *is* and what poetry *isn't* to be quite reductive. In part because they can be dismissive and exclusionary, and in part because poetry and language are constantly evolving. However, I accept that having some rough definitions or guidelines in place can at least give you the confidence that what you are writing really is a poem.

According to the Oxford English Dictionary the definition of a poem is a 'literary work in which the expression of feelings and ideas is given intensity by the use of distinctive style and rhythm.' This is actually quite an inclusive description, but please note that 'style and rhythm' are not an analogue for 'rhyme and metre'. I cannot state this clearly enough: poems do not have to rhyme, nor do they *have* to follow a strict metre (i.e. the basic rhythmic structure of a line within a poem or poetic work). This treatment of poetry is not a modern concept, as evidenced by the following quote from the English philosopher and political economist John Stuart Mill, published in an 1833 issue of the *Monthly Repository* [1, p. 60]:

> It has often been asked, What Is Poetry? And many and various are the answers which have been returned. The vulgarest of all – one with which no person possessed of the faculties to which poetry addresses itself can ever have been satisfied – is that which confounds poetry with metrical composition...

So, if poems don't have to rhyme or follow a strict metre, then what is it that defines them? Returning to the definition provided by the Oxford English Dictionary, the use of a distinctive 'style' implies a specific metre or structure, and while these forms (as mentioned already in Chap. 2) can provide a useful scaffolding for creativity, defining *all* poems as having to adhere to a set rhyme and/or metre is overly restrictive. However, a 'distinctive rhythm' is a much more generous and inclusive explanation of what a poem is. So let me propose the following definition for us to work with:

> All poems have rhythm.

Of course, some prose also has rhythm, and in terms of delineating poetry from prose, this is further complicated by the existence of both prose poems and poetic prose, but (at the risk of sounding trite) I honestly believe that you can tell if a poem is a poem by reading it and listening to its rhythm. To explore this idea a little further, let's look at the following two passages, both of which are written about a scientific study into safely reading sealed letters encrypted using letterlocking (a technique of folding a letter so that it becomes its own envelope) [2]. Both paragraphs have been written by me, and both are about the same subject, but which of them is a poem, and which of them is a piece of prose?

> Reams of dead letters hide correspondence beneath purposeful cuts and folds; the contents of their written past locked tight behind the paper-thin veneer of this faded piggy bank. Fold, tuck, slit, floss, roll, and then adhere. Wax, shear, route, starch, meanings disappear. Radiation probes delicately across the surface, creating cross structures to remind us of the

pains taken on behalf of senders now revealed; digitally unfurling history's creases to open secrets with seals intact.

In this study, a highly sensitive X-ray scanner, was used to scan a batch of unopened letters from a 17th-century postal trunk full of undelivered mail. The high sensitivity of the scanner (designed for intricately mapping the mineral content of teeth) made it possible to resolve certain types of ink in paper and parchment. Following the X-ray scanning of the letters, computational algorithms were applied to the scan images to identify and separate the different layers of the folded letter and 'virtually unfold' it.

Hopefully you managed to identify the first passage as the poem. While it is also true that there is a rather obvious rhyme of 'adhere/disappear' in this passage, it is not this which makes it a poem. Rather, it is the inherent rhythm of the piece. This poem does not have a specific metre or rhyming structure, and yet when you read it, you *know* that it is a poem.

Now I have a confession to make. I cheated slightly in this exercise. I stripped out the formatting of the poem, which as it was originally written looks like this:

Reams of dead letters
hide correspondence
beneath purposeful
cuts and folds;
the contents of
their written past
locked tight
behind the paper-thin
veneer of this
faded piggy bank.
Fold,
Tuck,
Slit,
Floss,
Roll,
And then adhere.
Wax,
Shear,
Route,
Starch,
Meanings disappear.
Radiation probes
delicately across the surface,
creating cross structures
to remind us of the pains
taken on behalf
of senders now revealed;

digitally unfurling
history's creases
to open secrets
with seals intact.

This version of the poem is structured via the use of different poetic techniques such as 'line breaks' (the termination of one line of poetry, and the beginning of a new line; represented elsewhere in this book by the use of '//'), 'enjambments' (the continuation of a sentence without a pause beyond the end of a line), and 'alliteration' (the repetition of the initial consonant sounds in words), which in turn act to make it *look* like a poem. These techniques also make the two versions of this poem different to read (try it), and no doubt they alter the rhythm of the piece. But crucially, removing the formatting does not r*emove* the rhythm.

Exercise 3.1: Play with Rhythm

Read this unformatted poem, written by me, about sea level rise that is killing trees along the Atlantic Coast, creating 'ghost forests' that are visible from space [3]. Where is the rhythm? Try to reformat it by adding in some line breaks so that it 'looks' more like some of the other poems that you have read so far in this book. How does this affect the rhythm? Does this reformatting affect the quality or the readability of the piece? Why?

Surging seas and weeping waves advance along your coast, probing buried channels as they break through the shoreface to drag briny fingerprints across weathered limbs that recoil at the touch. Tainted tides swell with pickled poison as saline sap pours down your brackish bark, below a crown of mottled grey that withers in the drink. Whisps of memories linger in brine, haunting faded shades of lost and broken greens.

Now have a look at the poem as I originally formatted it:

Surging seas and weeping waves
advance along your coast,
probing buried channels as they
break through the shoreface
to drag briny fingerprints
across weathered limbs
that recoil at the touch.
Tainted tides swell
with pickled poison
as saline sap pours down
your brackish bark,
below a crown of mottled grey
that withers in the drink.

> Whisps of memories
> linger in brine,
> haunting faded shades
> of lost and broken greens.

How does your version compare to this in terms of rhythm? Which do you prefer? Why?

As you work your way through the rest of this Chapter, developing your confidence in writing science poems, keep asking yourself about the rhythm of what you are writing. If what you have written has rhythm, then it is a poem. Have confidence in yourself as a poet, and do not let anyone else tell you otherwise.

3.3   Finding Inspiration

So, you want to write some science poetry, and you know that to write a poem it has to have rhythm. But what exactly are you going to write about?

To fully address this question, and to find inspiration for your science poem you should first consider two other questions: who are you writing for, and why are you writing for them? As discussed in Chap. 2, one of the most important steps for effective science communication (no matter what the medium) is to fully consider the specific audience you want to engage. In doing so you also need to consider why it is that you want to engage with them in the first instance. When writing your science poems, you fall into one of the four categories shown in Fig. 3.1, each of which I will now address in turn.

In Category 1 you know the audience you want to engage with your science poetry, but not the topic that you want to write about. For example, you might have been

Fig. 3.1 A graphical representation of how well you know your audience and scientific topic for a particular poem

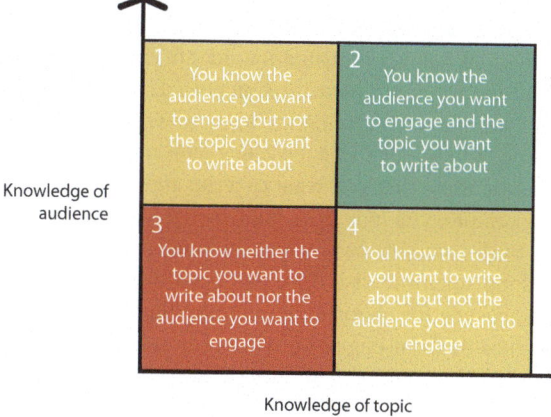

introduced to a local community group who are keen to become more involved in science, or you might be working with a nearby school to provide opportunities for their students to develop their knowledge beyond the taught curriculum. Whatever your situation, if you find yourself in this category then ask your audience what scientific topic they are interested in, and why. Doing so helps to build trust between you and your audience, and it also grants them agency, which in turn means that they are understandably far more likely to engage with you in the first instance [4]. Working with your audience to find the scientific topics that are relevant to them, rather than those that you *think* will be of interest will also ensure that you select the most appropriate scientific topics for effective engagement.

Imagine instead that you fall into Category 2, i.e. that you know which audience you want to engage with your poetry and the topic that you want to write about for this audience. For example, you might be a volcanologist who wants to communicate with a community of indigenous peoples who are living near to a potentially active volcano because you want to exchange knowledge with regards to signs of imminent volcanic activity [5]. Or else you might be a science communicator who wants to challenge the western hegemony of astronomy and use this to develop a conversation with schoolchildren about different sky stories and cultures [6]. In both examples there is a specific audience to write for, a specific reason for engaging with this audience, and a specific topic to write about, informed by the needs of the audience. Despite this supposed certainty however, I would still advise asking your audience about the scientific topic in question and what it is about this topic that particularly resonates with them, or not. Doing so will help to ensure that your science poem(s) are more effective at engaging with this audience.

Suppose that you belong in Category 3—you want to write some science poetry, but you know neither your intended audience nor your topic. The first thing you need to do is to find your audience. Who do you want to engage and why? This could be an audience that you know is particularly underserved by your scientific field, or by science communication more generally; but it could equally be a friend, family, or colleague. Alternatively, you might consider engaging with an audience that is already familiar to you (for example your followers across various social media channels), or else you might be interested in science-themed submissions to poetry journals (see Sect. 3.5). You might even want to write a poem specifically for yourself. Whatever the situation, identify and engage with your audience, and then repeat the steps outlined for Category 1, as you have now found an audience, but don't yet know what scientific topic to write about.

And finally, let's conceive that you are in Category 4, i.e. that you know the scientific topic that you want to write about but are unsure who you are writing for. For example, you may have published a new study, the findings of which you would like to disseminate to a wider audience, or you may have heard about a ground-breaking scientific discovery and been inspired to write a poem about it. Whatever your reasons for choosing a scientific topic, you should now find an audience who will benefit from engaging with your subsequent poem. If you already engage in other science communication activities and have an audience with whom you engage (e.g. followers of your blog or Twitter account) then you could write the poem for them. Alternatively, you should think about finding an audience who would want to engage with a science poem on this topic; for example, if you are writing about a newly discovered comet is there a group of amateur astronomers who might value your contribution? Similarly, you might find that your poem would appeal to other scientists working in the field of your chosen topic, or decide that you want to write about this topic because it appeals to you (and thus you are the intended audience of the poem). Once you succeed in finding an audience that relates to your chosen scientific topic, I would again advise engaging with them in the first instance to find out what it is about this specific topic that particularly appeals to them. Doing so will not only make them more likely to engage with your poem when it is finished, but it will also present you with ideas and imagery that you can subsequently weave into the construction of the poem itself.

Exercise 3.2: Find Your Audience
Imagine that you are in Category 3 of Fig. 3.1, i.e. that you know neither your intended audience nor your topic for a science poem. This audience could be your partner, your best friend, one of your colleagues, or even yourself. Once you have identified an audience, collaborate with them to find a scientific topic that is of interest to them, taking note of why they find it to be so engaging. You'll be using this information to help you to write some science poems in the next set of exercises in this Chapter.

Whichever of the four categories in Fig. 3.1 you find yourself in (and this is unlikely to remain consistent), the most important thing when you are considering what to write your science poetry about is to think of your audience. Who are your audience, and why do you want to engage them? What scientific topics are relevant and pertinent to them, and how will writing science poetry help them to engage with these topics?

3.4 Writing Science Poetry

You have your audience. You have your topic. So, what do you do now? Well, you need to find your rhythm.

As discussed in Chap. 2, when writing science poetry, it really helps to start with a specific poetic form. Not because poetry *must* follow such a form (See Sect. 3.2), but rather because following a form will really help you to develop your skills in writing poetry. Once you have practiced using a few forms and have begun to build your confidence as a poet, then you can break away (and even break up) these forms, but for now they are a great place for you to learn you craft. Some of these forms also offer an effective way to further reinforce the science that you are trying to communicate, through a combination of deliberate repetition, structure, and/or shape.

In the rest of this Section, I am going to introduce you to three different types of poetic form, providing examples and exercises to help you to create your own science poems. First up, the 'kyrielle'.

The kyrielle is a French form of rhyming poetry that is written in blocks of four lines. The technical term for one of these blocks is a 'stanza', and stanzas that contain four lines are known as 'quatrains'. A kyrielle can have any number of these quatrains, but three is the expected minimum, and the fourth line of each quatrain is always the same; this is what is referred to as the 'refrain' (i.e. a line that is repeated).

Each line of the quatrain should be exactly eight syllables, and tends to adopt one of the two following rhyming patterns: aabB ccbB ddbB and abaB cbcB dbdB. This is standard nomenclature for rhyming patterns, where matching lowercase letters indicate that the last syllable in these lines is (more or less) the same and the uppercase 'B' represents the kyrielle's refrain (in this nomenclature capital letters indicate a repetition of the whole line rather than just a matching syllable). The structure of a kyrielle is illustrated in Fig. 3.2.

To see what this looks like in practice, here is a kyrielle of mine entitled 'Learning to Listen', which follows an aabB ccbB ddbB eebB ffbB ggbB rhyming pattern:

As weathers change the birds migrate,
Flocking en masse like living freight;
They have decanted overseas,
The notes of birdsong on the breeze.

In mapping out when this takes place,
We listen for a vocal trace;
And sift through sounds with expertise,
The notes of birdsong on the breeze.

Fig. 3.2 Two different structures for a kyrielle

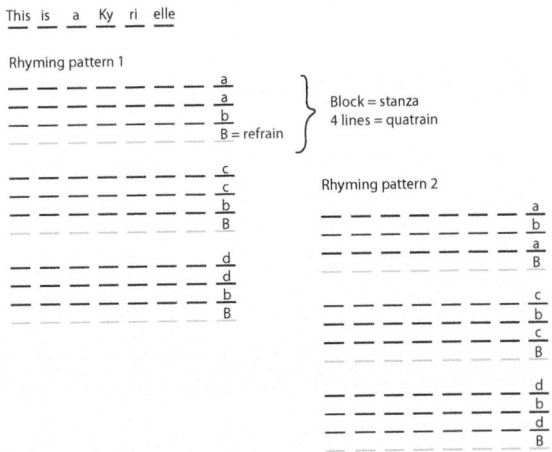

But straining every avian whine
Requires precious thought and time;
Through manmade sounds we wish to tease,
The notes of birdsong on the breeze.

Computers can be trained to sort
The songbirds from the road transport;
They pick them out with perfect ease,
The notes of birdsong on the breeze.

So now that they are trained to find,
Migration shapes can be refined;
But changing forms might not appease,
The notes of birdsong on the breeze.

As climates change the birds migrate,
Machines map out their distant fate;
An empty screen appears to freeze,
The notes of birdsong on the breeze.

This poem was inspired by research which found that machine learning can be used to better determine bird migration patterns, and the effects that climate change is having on them [7]. You can see that in this instance the strict rhyming pattern helps to reflect the algorithmic nature of the underlying research, while the refrain has an almost musical quality to it which is reminiscent of the notes of birdsong that it is trying to replicate. In this instance, the structure of the poem reinforces the narrative of the scientific research that it is trying to communicate.

Exercise 3.3: Write a Kyrielle
Using the scientific topic that you identified in Exercise 3.2, and keeping your target audience in mind, write a five-quatrain kyrielle using the aabB ccbB ddbB eebB ffbB rhyming scheme. When you are writing a kyrielle the first thing that you should do is to determine what the refrain will be. This should be impactful, and relevant to the scientific topic that you are representing. It also needs to have lots of possible rhyming combinations, so choose carefully! I find that rhyming dictionaries (there are many available for free online) are extremely useful for this, and indeed any other, form of rhyming poem.

The next poetic form that I would like to introduce you to is the 'Shakespearean sonnet'. There are several different forms of sonnet, but the Shakespearean (or English) sonnet is perhaps the best known, made famous by The Bard from whom it takes its name. This form of sonnet consists of 14 lines, made up of three quatrains and a couplet (a pair of consecutive lines), with the following rhyming scheme: abab cdcd efef gg. Each of these lines is also written in what is known as 'iambic pentameter'.

Poetic metre is typically split up into units called 'feet', and an 'iamb' is a foot for which there is two syllables, the first of which is unstressed and the second of which is stressed. For example, the words 'refine', 'later', and 'attain' are all examples of this iambic pattern of an unstressed syllable followed by a stressed one. Pentameter just signifies that there are five feet in each line of the poem, so iambic pentameter means that each line has 10 syllables, following an 'unstressed/stressed/unstressed/stressed' pattern. The following lines are all examples of iambic pentameter: 'the rain in Spain falls mainly on the plane', 'now is the winter of our discontent', and 'come on and slam and welcome to the jam'.

In addition, the Shakespearean sonnet has what is known as a 'volta' (the Italian word for a 'turn'), in which the rhyme scheme and the subject of the poem suddenly change, introducing a different viewpoint (or juxtaposition) which gives new meaning to the ideas that have already been expressed in the poem. In a Shakespearean sonnet this volta happens after the third quatrain and before the final couplet (i.e. between the twelfth and thirteenth line). The formatting of a Shakespearean sonnet is shown in Fig. 3.3.

To illustrate this form, read the following poem 'An Astronaut's Broken Heart', written about research which found that astronauts who were part of the Apollo missions (who travelled beyond the Earth's magnetosphere, where they were

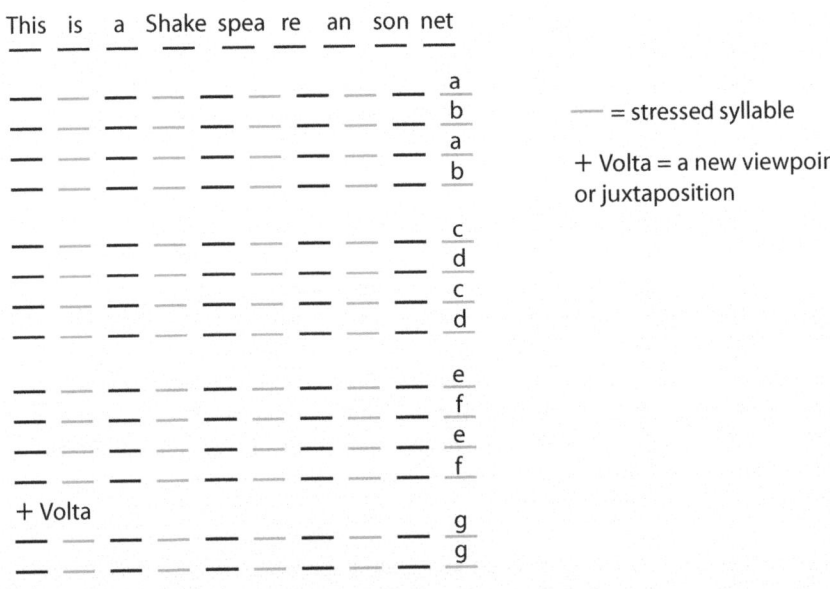

Fig. 3.3   The structure of a Shakespearean sonnet

subjected to large amounts of deep space radiation) were almost five times more likely to die from cardiovascular diseases than other astronauts [8].

> Launched into space like puppets on a string,
> These metal ships danced fiercely on the waves;
> Such precious cargo was held deep within,
> Laden with men who would see early graves.
> A tranquil sea lay out before their eyes,
> Revealing scenes that only they have seen;
> The shadow of the Earth in a sunrise,
> Would fade and slowly die like a lost dream.
> Now far from Gaia's firm, protective reach,
> Their bodies throbbed with substances unknown;
> And somewhere in their minds a telling breach:
> When you have seen the Earth is it still home?
> Should deep space radiation take the blame,
> When living could not ever be the same.

In this example you can see that the poem follows the strict rhyming pattern of a Shakespearian sonnet, and that each line is written in iambic pentameter. I have also used the volta (coming after the line 'When you have seen the Earth is it still home?') to question if there might be another reason for why these astronauts were more likely to die from cardiovascular diseases, i.e. because travelling so far from Earth might have made such a return more difficult on an emotional level.

Exercise 3.4: Write a Shakespearean Sonnet

Using the same scientific topic as Exercise 3.3, write a Shakespearean sonnet. Pay attention to the iambic pentameter, and to the volta. Think about what your turn will be and how this will help the reader to recontextualise the lines that precede it. As with the kyrielle, be careful when choosing your line endings: pick words for which you can find a suitable number of rhymes.

The third, and final, poetic form that I would like to introduce is the 'pantoum'. The pantoum is a Malaysian form of poetry that features repeating lines. They are composed of any number of quatrains in which the second and fourth lines of one serve as the first and third lines of the next, with each line typically between 8 and 12 syllables in length. The last line of a pantoum is also usually the same as the first, with each quatrain following an ABA'B' rhyming scheme and a new rhyme introduced with every quatrain (here the ' is used to indicate that the lines A and A' are different from each other, but that they rhyme and also remain the same throughout the poem). This means that a four-stanza pantoum would have the following rhyming structure: ABA'B' BCB'C' CDC'D' DA'D'A. Unlike the kyrielle, in which the refrain is often used to reinforce a specific message, the repeating lines in a pantoum usually change meaning with context, which lends this form of poetry extremely well to communicating science that is related to transformation or progression. This is quite a complex form of poetry, but Fig. 3.4 should help you to visualise it better.

The following poem, 'The Sound of Shapes' is an example of a pantoum. This was written by me as a response to research that uncovered molecular clues into the potential causes of synaesthesia, a condition in which two or more of the five senses that are normally experienced separately are instead experienced simultaneously (e.g. seeing colours while listening to music) [9].

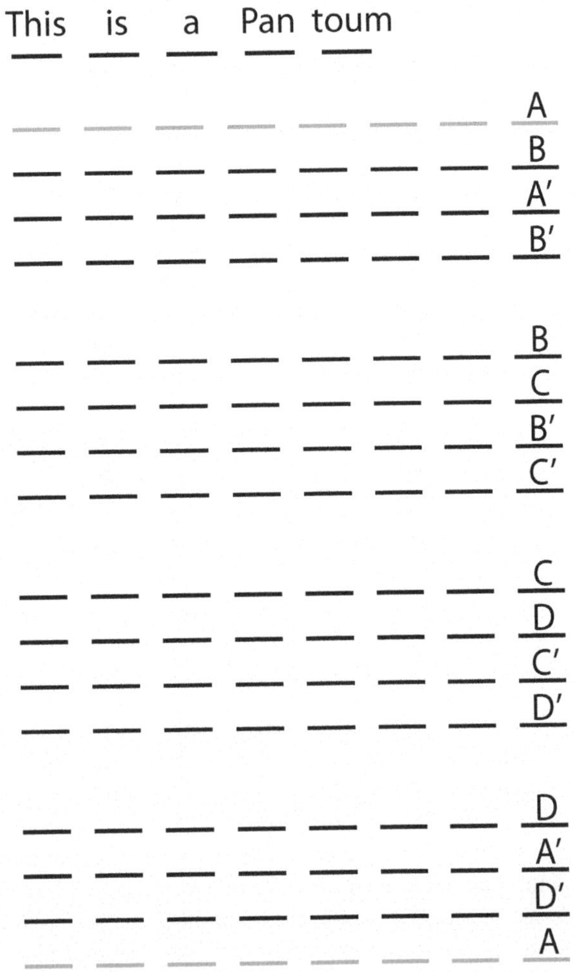

Fig. 3.4 The structure of a pantoum

The soothing sound of blue floats past my sight,
I open up my eyes to hear a shape;
The colours and the music are not right,
The painting tastes more like a stark landscape.

I open up my eyes to hear a shape,
As Turner's clouds begin to smell of damp;
The painting tastes more like a stark landscape,
With senses overwhelmed until they cramp.

As Turner's clouds begin to smell of damp,
My mother says they sound like rippled spray;
With senses overwhelmed until they cramp,
The petals falling sound like a Monet.

My mother says they sound like rippled spray,
As neurons in my brain begin to beat;
The petals falling sound like a Monet,
Accompanied by the taste of something sweet.

As neurons in my brain begin to beat,
The colours and the music are not right;
Accompanied by the taste of something sweet,
The soothing sound of blue floats past my sight.

Here the pantoum is especially effective at helping to embody the bewilderment that is often associated with synaesthesia, with the differing meanings of the repeated lines making the reader feel simultaneously assured and uncertain.

Exercise 3.5: Write a Pantoum
Using the same scientific topic as exercises 3.3 and 3.4, write a four-quatrain pantoum, which follows the rhyming pattern of ABA'B' BCB'C' CDC'D' DA'D'A. Begin by picking you first line very carefully, as this will also form the final line of your poem, framing the way in which it is perceived and interpreted by the reader. With each new line that you write, remember also that it will appear again, so try and give it flexibility for multiple interpretations.

The kyrielle, the Shakespearean sonnet, and the pantoum are three versatile and relatively short forms of poetry that will help to build your confidence as a writer, providing you with the scaffolding that you need to develop your rhythm as a poet. There are many other poetic forms that you might try, including (but not limited to): the 'villanelle', the 'ghazal', the 'terza rima', and the 'sestina'. Each of these different forms has their own unique patterns and structures to comprehend, and while some of them may feel constricting, learning how they work and the rhythms that they produce is like learning how to play scales with a musical instrument, doing so gives you the underlying knowledge and confidence to develop your creativity. Furthermore (as with the other forms of poetry that have appeared so far in this book), each form has specific traits that may lend themselves to communicating a specific scientific topic more (or less) effectively.

Speaking from personal experience, I spent many years learning how to write science poems based entirely on established poetic forms, and while I now tend to write in 'free verse', using these forms gave me confidence and helped me learn how to observe rhythm in both my writing, and in my day-to-day life. As while it is true that free verse does not use consistent metre or rhyme, it instead tends to follow the rhythm of natural speech. This means that writing effective free verse can be exceptionally difficult, unless you have a very good feeling for rhythms and how they work. Working with poetic forms is the best way that I have found to build an innate sense of rhythm, and instead of being constrictive it has certainly helped me to develop my own unique poetic voice.

As you start to find your own rhythm, you might also want to consider using different poetic techniques such as line breaks, enjambments, alliteration, and 'caesuras' (the use of a pause, or breath, mid-line). The best explainer I have come across for the practical implementation of these techniques is Stephen Fry's *The Ode Less Travelled* [10]. As with the use of poetic forms, these techniques are not mandatory, but learning what they are, and how you might implement them in your writing, is a really effective way to develop your craft.

After you have written your science poem you need to edit it. It is highly unlikely that the first version of the poem that you have written is the best version that it can be. You need to give the poem some space to breathe. Leave it percolating for a couple of days, or at the very least step away from your notepad or computer for a short break. When you come back to the poem with fresh eyes, read it out loud and in your head, and see what you think now. How is the rhythm? Does any of it sound stilted? Are there any lines which are less good than the others, or equally are there any lines which are brilliant, but which don't quite fit the rhythm of this piece? If so, then cut them out and save them for another poem. I would recommend doing this at least two or three different times, and then showing your poem to a trusted confidant, ideally a member of your intended audience, who can be relied upon for honest and respectful feedback. Take their critique on board and treat it like you would the peer review for a scientific article: you don't have to agree with all of their comments or recommended changes, but you should be able to defend your decisions.

A temptation when writing science poetry is to be fairly didactic, i.e. poetry in which the scientific research is communicated directly to the intended audience (e.g. 'How much did the pH rise in the sea? // By a statistical factor of almost 2.93'). However, the best science poetry is that which engages its audience, while also encouraging them to want to find out more about the science that is being discussed in the poem. This is a skill that comes with practice, mainly through learning which are the most engaging parts of the scientific topic for your audience and working out ways that you can use your writing to elicit such engagement. As with editing your work, working with your target audience to determine the extent to which you have managed to do this is key to your development as a poet.

The final topic that I would like to cover briefly in this section on writing science poems is how to find a suitable title. Deciding on the best title for your poems is really hard (or at least I find it to be hard). The best advice I have received on this topic was from the poet Sara Goudarzi, who told me that 'the title should say something that the poem does not.' How I interpret this when writing my own science poems, is that if the scientific topic is immediately obvious from the poem itself then I can afford to be a little bit more playful with the title, but if the poem is deliberately vague or metaphorical then a more literal title might be useful for my intended audience. To do this effectively I would recommend working out your title only after you have written your poem, and to try and avoid simply using one of the lines from the poem itself, tempting as this might be.

3.5 Sharing Your Poems

Now that you have written your science poem, edited it, worked out a title, edited it some more, and then some more, what do you do with it? If you've sequentially worked your way through the various sections in this Chapter, then you will have already identified an audience for your poetry, so begin by sharing it with them.

As discussed in Sect. 3.4, you might already have shared it with some members of this audience to elicit feedback from them during the editing stage. Having now finished your poem, you need to share it with your target audience and see what reaction it provokes (Fig. 3.5 summarises this approach for writing and sharing a science poem). Does it disseminate knowledge or engender dialogue in the manner which you imagined, and did you receive any further feedback (solicited or not) that you could use to make another edit of your poem?

Imagine, for example, that you have written a science poem to raise awareness and initiate dialogue with local landowners about radon and the impact that this might have on building regulations for new homes in their area [11]. With a clearly defined audience, scientific topic, and reason for engagement, suppose that you got in touch with some members of this audience, asked them about their needs and experiences, and then wrote a science poem. You might even have shown an earlier version of your poem to a couple of local landowners and responded to their feedback. So how do you now reach all of your intended audience?

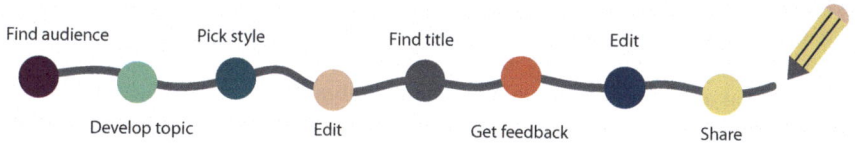

Fig. 3.5 The process for writing and sharing a science poem

Begin by working with your contacts in the community to identify the media through which they themselves communicate. Do they have a newsletter, mailing list, or social media presence (e.g. a Facebook page or Twitter account) that you could use to share your poem and solicit feedback? Working with your intended audience to distribute your poem across that community is by far the most effective way of ensuring that it is seen, giving you the best chance to capture the extent to which it is engaged with.

What if you want to reach an even wider audience? Suppose that instead of writing a science poem to address a specific scientific topic with a relatively narrow audience you instead want to raise awareness of a piece of scientific research to a potentially much larger audience. The advice from Sect. 3.3 still holds true; you need to begin by thinking about which specific audience you want to target and why. For example, do you want to use a science poem to communicate the risks of air pollution to pregnant women [12] or to start a dialogue about the dangers of overfishing and climate change [13] with the fishing industry. If so, then what is the most effective way to reach these audiences via their chosen forms of media?

Even if you want to think in the absolute broadest of terms (e.g. using science poetry to communicate an aspect of your own scientific research to anyone who is not a specialist in the field), you should avoid the mistake of trying to target a 'general' public, and instead break down your approach into targeting several different publics. For example, imagine that you are a researcher working on the impact of ocean acidification on coral reefs [14], and that you want to write a science poem to raise awareness of ocean acidification amongst people who might not be aware what it is and why it is important. This is potentially a very large audience, which in turn is made up of many publics (none of whom primarily identify themselves as 'non-scientists'). Instead of trying to reach all of this audience through one medium (e.g. a blog, Instagram post, podcast) think about some of the different publics that make up this larger audience (e.g. school children in a coastal community, policymakers, journalists), how you would like to engage them, for what purpose, and through which media.

Depending on your target audience, social media can be a particularly effective way of reaching many people. As well as containing various communities which might act to serve as your intended audience, Instagram, Twitter, Reddit, and Facebook are all well-established vessels through which poetry is regularly shared and curated. Furthermore, these platforms can also provide useful communities through which to get feedback on your science poetry, helping you in the editing process and developing your skills as a writer. As with any feedback, be sure that you carefully outline the parameters within which you want to receive it, and don't feel obliged to incorporate any of it.

Another strategy for sharing your poetry is to submit it to some poetry journals or to enter it into some competitions. In this instance your audience would be the readers of the journal and/or organisation that you are submitting your work to. However, before doing so it is essential that you read some of the poetry that has previously been published there, to make sure that your work aligns.

In Chap. 2, I mentioned several poetry journals that have a scientific focus, but there are hundreds (if not thousands) of poetry journals and poetry competitions that you might submit your science poems to, providing that they meet the conditions of submission. *Submittable* [15] is an online submission management system that can enable you to keep track of the poems that you submit, as well as providing an updated list of any upcoming submission opportunities and the deadlines and guiding principles for each.

Reading poetry from these journals will not only help you to better understand the kind of submissions that will resonate with their editors / judges and engage their audiences, but it will also help to make you a better poet. Just as keeping up to date with the scientific literature in any field is essential for better understanding opportunities for development, discovery, and communication, so too will reading poetry help you to better comprehend how different poetic techniques are used in practice, while helping you to develop your own unique and discernible voice as a poet.

Exercise 3.6: Find Some Competitions

Create a free account with *Submittable* and do a search for any upcoming poetry submissions. Have a look at some of these opportunities and depending on your budget (many of these submissions charge a fee to help the journal or organisation with their running costs) and availability, write a poem and submit it. You might use one of the poems that you have written in this Chapter, but if you do then make sure that it adheres to both the strict rules for submission (e.g. line count, theme) and also the general ethos of the journal or organisation to which you are submitting. The best way to do this is to read some of the poetry that has previously been submitted there. Doing so will also help to support their work, as well as that of other poets.

3.6 Performing Poetry

You've written a poem, you've edited it, decided that it is in fact a poem, and have even shared it with your target audience. So, what now? Well, you might want to perform it. Before I discuss different ways in which you could do this, and the benefits for doing so, I just want to make two things clear. Firstly, spoken word poetry is different from written poetry, and what makes for an impactful and engaging spoken word piece does not always translate as effectively onto the written page, and vice versa. And secondly, written poetry does not have to be performed; the final version of any science poem that you have created can exist entirely on the written page. It does not have to be read out loud by you or anyone else to somehow make it 'better'.

What are the benefits for performing your poetry, and how might you go about doing so? Sharing your science poetry as a piece of spoken word can help you develop your rhythm and confidence as a poet, and it can also help you to engage with audiences that you might not ordinarily reach with written versions of your work.

Perhaps the most straightforward way to perform and share spoken word poetry is to make an audio recording of your poem and then host and share it via a social media site that specialises in audio files, such as SoundCloud. You can record your poem using the microphone on your smartphone or laptop and then edit it using a range of editing software (I would recommend the cross-platform, open-source digital audio editor Audacity). Reading your poem out loud to yourself during this process will also help you to hone its rhythm, although as stated above this might be quite different from the way in which it appears on the page (and when it is read in your head).

You might also want to share your poetry via a podcast, especially if you know that this is a medium that is used by the audience(s) whom you wish to engage. There is plenty of advice written elsewhere about how to create effective science communication content for podcasts [see e.g. 16], but begin by thinking about which podcasts your intended audience might already listen to and then reach out to the creators of these podcasts to see if they would be interested in featuring your science poem.

The two other main ways in which you might engage audiences with your science poetry are via open mics and poetry slams. These are both poetry events that feature a selection of poets reading their work out loud to an audience, although they are quite different in their setup and execution.

Open mics tend to be events that are organised for poets to share their poetry in an inclusive and collaborative environment. Most open mic events will feature one or two more-established poets performing a set (or selection) of their poems, interspersed by segments in which other attendees are welcome to step up to the microphone (mic) and read their poems. It is usual for poets to sign up to read their poems either in advance or at the beginning of the event, and it is accepted that some of these poems will be works in progress, or that they might be delivered by poets who are still building confidence in sharing their work in a public arena.

Open mic events will generally be hosted by a compere (or emcee), who will make sure that the audience is respectful to the poets, helping to create a safe space for sharing their work. At least that is how open mic events work in theory. Some comperes are better than others at making these events accessible and inclusive, and so it is recommended to attend a few such events so that you know what to expect before sharing your own science poems. There will also usually be a time limit per poem (normally 2 or 3 min), and it is considered bad form to simply read your poem and then leave. Listening to the other poets will also help to develop your own performance and delivery style, just as reading the work of other poets will help you to develop your writing.

Poetry slams differ from open mic events in that they are primarily run as competitions. Each poet competes against the other poets to be crowned the 'winner'. While every poetry slam is run slightly differently, a typical format is for each poet to take it in turn to read their poems (normally one or two per poet) after which they will be judged and scored against a set of criteria (delivery, content, etc.) by a panel of judges and/or the audience. Some poetry slams have multiple rounds (in which the lowest scoring poets drop out at each stage), whereas others have several qualifying events with the winner of each making it through to a final. As with an open mic event, poetry slams are hosted by an emcee, and while part of their role is to make the competing poets feel comfortable, the competitive nature of poetry slams means that they can be more intimidating to perform at than open mics.

Poetry slams are a great way of learning how to develop your own performance and delivery, but there is also a tendency for them to be dominated by a very particular type of poem and delivery. While this is not always the case, it is worth attending several different poetry slams before entering one for yourself. Most poetry slams also tend to begin with a 'sacrificial poet'. This is a poet who delivers their poem according to the rules of the competition, and is judged according to that slam's criteria, but solely for the purpose of allowing the judges to calibrate their judging. The sacrificial poet does not enter the slam itself and tends to get an easier ride from both the judges and the audience, making it a good opportunity to test the water when you think you are ready to enter a poetry slam.

Exercise 3.7: Listen to Some Science Poems
There are many great examples of spoken word poetry on the Internet, and SoundCloud in particular contains thousands of science poems, across a wide range of voices and poets. Go to SoundCloud and perform a keyword search for the words 'science' AND 'poetry'. Listen to some of these poems and consider what it is about their performance that makes them particularly effective or not. If the poet has provided a link to the written version of the piece read this as well and see how you might deliver it differently. Finally, ask yourself if you prefer the poem as it is written on the page or when it is spoken out loud, and consider why you think this is the case.

3.7 Summary

This Chapter has built on the ideas that were introduced in Chap. 2 to develop your skills in writing science poetry. After reading this Chapter you should have a better idea of why you want to write science poems, the audience who you want to write for, and the scientific topic that you want to engage with. You should also have a clearer understanding of what makes a poem (rhythm), and how you can use various poetic forms to help build your confidence as a writer. This Chapter has also provided practical advice and guidance for how you might share your science poetry with a wider audience, including poetry journals and competitions.

Finally, a primer to performing poetry has been presented, alongside a rationale for why you might want to take your poetry from the page to the stage. Many of the exercises that have been presented in this Chapter can also be used if you decide to develop your own science poetry workshops and activities (see Chap. 6), and by working through them yourself you will gain confidence in finding your rhythm and developing your own unique poetic voice for the communication of science.

3.8   Suggested Reading

As I have already mentioned several times now, Stephen Fry's *The Ode Less Travelled* [10] is an original and accessible guide to poetic form, metre, and technique. *Writing Poems* by Peter Sansom [17] provides an equally useful reference guide on how to develop your confidence as a poet, including brief introductions to different poets and the poetic styles they developed. For those wishing to develop the performance aspect of their poetry *Take the Mic: The Art of Performance Poetry, Slam, and the Spoken Word* [18] provides an authoritative account, and was co-written by Marc Smith, who began the poetry slam movement in the 1980s in Chicago. Finally, the work of Michael Dahlstrom [19], Carol Rogers [20], and Emily Dawson [21], as well as Matthew Nisbett and Dietram Scheufele [22] are essential for understanding the role that the audience can (and should play) in scientific discourse and how we can start to engage these audiences to help their voices to be both heard and actioned.

3.9   Further Study

The further study in this Chapter is designed to help you develop your rhythm as a scientific poet, to consider your audience, and to think about how you might share your work in a public space.

1. **Try some more poetic forms**. The website *Shadow Poetry* [23] provides a comprehensive list of many different types of poetic form, including several examples for each. Pick two poetic forms and write two new poems about the same scientific topic that you used for your kyrielle, Shakespearean sonnet, and pantoum earlier in this Chapter. If you really want a challenge, then try writing a sestina.
2. **Find a new audience**. Think about the audience(s) that you identified in Sect. 3.3 for your science poem(s). Are there some other audiences you could work with, and with whom you have not previously engaged? What would be the challenges and benefits for engaging this audience, and how would you identify the scientific topics (and poetic forms) that were of interest to them?

3. **Sign up to an open mic**. Find an open mic event that is happening either in your local area or in a virtual environment and sign up to read one of the science poems that you have written in this Chapter. Before you do so, go along to a couple of these events and make sure that the emcee creates an environment in which the poets are respected, and where it is clear that you will feel safe and comfortable in giving a reading.

References

1. Mill JS (1833) What is poetry? Mon Repos 7(73):60–70
2. Dambrogio J, Ghassaei A, Smith DS et al (2021) Unlocking history through automated virtual unfolding of sealed documents imaged by X-ray microtomography. Nat Commun 12(1):1–10. https://doi.org/10.1038/s41467-021-21326-w
3. Ury EA, Yang X, Wright JP et al (2021) Rapid deforestation of a coastal landscape driven by sea level rise and extreme events. Ecol Appl 31(5):e02339. https://doi.org/10.1002/eap.2339
4. Roberson T (2020) On social change, agency, and public interest: what can science communication learn from public relations? J Sci Commun 19(2):Y01. https://doi.org/10.22323/2.190 20401
5. Kelman I, Mercer J, Gaillard JC (2012) Indigenous knowledge and disaster risk reduction. Geography 97(1):12–21. https://doi.org/10.1080/00167487.2012.12094332
6. Ruddell N, Danaia L, McKinnon D (2016) indigenous sky stories: reframing how we introduce primary school students to astronomy—a Type II case study of implementation. Aust J Indig Educ 45(2):170–180. https://doi.org/10.1017/jie.2016.21
7. Oliver RY, Ellis DPW, Chmura HE et al (2018) Eavesdropping on the arctic: automated bioacoustics reveal dynamics in songbird breeding phenology. Sci Adv 4(6):eaaq1084. https://doi.org/10.1126/sciadv.aaq1084
8. Delp MD, Charvat JM, Cl L et al (2016) Apollo lunar astronauts show higher cardiovascular disease mortality: possible deep space radiation effects on the vascular endothelium. Sci Rep 6:29901. https://doi.org/10.1038/srep29901
9. Tilot AK, Kucera KS, Vino A et al (2018) Rare variants in axonogenesis genes connect three families with sound–color synesthesia. Proc Natl Acad Sci 115(12):3168–3173. https://doi.org/10.1073/pnas.1715492115
10. Fry S (2007) The ode less travelled: unlocking the poet within. Arrow Books, London
11. Denman AR, Crockett RG, Groves-Kirkby CJ (2018) An assessment of the effectiveness of UK building regulations for new homes in radon affected areas. J Environ Radioact 192:166–171. https://doi.org/10.1016/j.jenvrad.2018.06.017
12. Bai W, Li Y, Niu Y et al (2020) Association between ambient air pollution and pregnancy complications: a systematic review and meta-analysis of cohort studies. Environ Res 185:109471. https://doi.org/10.1016/j.envres.2020.109471
13. Schartup AT, Thackray CP, Qureshi A et al (2019) Climate change and overfishing increase neurotoxicant in marine predators. Nature 572:648–650. https://doi.org/10.1038/s41586-019-1468-9
14. Andersson AJ, Gledhill D (2013) Ocean acidification and coral reefs: effects on breakdown, dissolution, and net ecosystem calcification. Ann Rev Mar Sci 5:321–348. https://doi.org/10.1146/annurev-marine-121211-172241
15. Submittable (2021) Submittable 2021. https://www.submittable.com. Accessed 10 Dec 2021
16. Illingworth S, Allen G (2020) Effective science communication, 2nd edn. Institute of Physics Publishing, Bristol
17. Sansom P (1994) Writing poems. Bloodaxe Books Ltd., Highgreen

18. Smith MK, Kraynak J (2009) Take the mic: the art of performance poetry, slam, and the spoken word. Sourcebooks Inc., Naperville

19. Dahlstrom MF (2014) Using narratives and storytelling to communicate science with nonexpert audiences. Proc Natl Acad Sci 111:13614–13620. https://doi.org/10.1073/pnas.1320645111

20. Rogers CL (2000) Making the audience a key participant in the science communication process. Sci Eng Ethics 6(4):553–557. https://doi.org/10.1007/s11948-000-0015-1

21. Dawson E (2018) Reimagining publics and (non) participation: Exploring exclusion from science communication through the experiences of low-income, minority ethnic groups. Public Underst Sci 27(7):772–786. https://doi.org/10.1177/0963662517750072

22. Nisbet MC, Scheufele DA (2009) What's next for science communication? promising directions and lingering distractions. Am J Bot 96(10):1767–1778. https://doi.org/10.3732/ajb.0900041

23. Shadow Poetry (2021) Shadow poetry. http://www.shadowpoetry.com. Accessed 10 Dec 2021

Chapter 4
Poetic Content Analysis

4.1 Introduction

So far in this book I have discussed how to read, analyse, and write science poetry so that you can start to use it as a medium through which to communicate science to different audiences. In Chaps. 4 and 5 I will go on to introduce two forms of poetic inquiry, demonstrating how poetry can be used as the basis for two distinct research methods, the purposes of which are to interrogate both science and scientific discourse. In Chap. 4 I will describe how poetry can be used as a form of qualitative content analysis, and in Chap. 5 I will go on to discuss poetic transcription. In short, the differences between these two research methods are that poetic content analysis involves analysing poetry written about a specific topic or theme, whereas poetic transcription involves creating poetry from other qualitative data (e.g. interviews, field notes, survey responses).

Before I go on to discuss qualitative content analysis, and the poetic turn that I have developed for this research method, it is important to make the following distinction between a research method and a research methodology: a method is the research tool (e.g. interview, questionnaire, observation) that is used by a researcher, and a methodology is the justification for using this method. To justify the poetic content analysis method that I lay out in the remainder of this Chapter, it is thus first necessary to consider the theoretical perspective, and in turn the ontological and epistemological stances of this approach.

Ontology is concerned with what kinds of things exist in the world, while episte-mology deals with how we can know what exists [1]. Your ontological (what exists) and epistemological (how you know) perspectives shape not only your attitudes as a researcher and science communicator, but also the ways in which you interpret the natural and social worlds around you. Your ontological and epistemological posi-tions are driven by your theoretical perspective, and broadly speaking there are two of these to consider: positivism and interpretivism. A positivist mindset is one that supposes a single external reality (ontological perspective) that it is possible to know objectively (epistemological perspective), whereas an interpretivist mindset is one

S. Illingworth, *Science Communication Through Poetry*,
https://doi.org/10.1007/978-3-030-96829-8_4

that believes there to be no single reality (ontological perspective) and that trying to make sense of this is a subjective process that depends on how it is interpreted (epistemological perspective).

In reality there are more than just these two distinct theoretical perspectives, and alongside several different 'flavours' of interpretivism (e.g. social constructivism and phenomenology) there are also several other stances such as critical inquiry, postmodernism, and feminism to be considered [2]. However, the purpose of this brief introduction is to emphasise the importance of choosing a research method that is congruent with both the way in which you view the world, and how you believe science should be communicated. For example, if you adopt a positivist mindset then a focus group (an example of a qualitative research method) would be an odd way of trying to obtain objective knowledge in relation to an absolute reality. Similarly, a purely statistical analysis (an example of a quantitative research method) is not entirely harmonious with an interpretivist perspective.

As I outlined in Chap. 2. I don't believe that there is any single 'true' meaning of a poem, i.e. poetry is not an objective certainty, and its interpretation is largely subjective. It therefore follows that the method I now present is underpinned by an interpretivist perspective—it does not seek to find a singular truth, but rather offers a method for interpreting the different ways in which science is itself perceived.

4.2 Qualitative Content Analysis

Content analysis is a set of qualitative research methods that are used for identifying and analysing patterns of meaning in a data set [3]. Rather than being a singular tool, there are generally three distinct approaches that can be adopted: conventional, directed, and summative [4]. A conventional approach involves analysing the data and observing what codes (or labels) emerge, whereas a directed approach involves pre-determining a series of codes according to an initial theory or research question. In contrast to these two approaches, a summative content analysis is more quantitative in nature and generally involves counting the frequency of specific keywords or phrases to explore their usage.

The choice of content analysis that is employed by a researcher depends on their theoretical perspective and research methodology, and as discussed in Sect. 4.1 it is essential that the most congruent method is chosen to ensure the most appropriate analysis. While a directed and/or purely summative approach to content analysis has previously been used to analyse poetry [see e.g. 5, 6] the subjective nature of poems makes them more suited to a conventional approach, in which the codes, categories, and themes (see below) emerge from the process itself.

Conventional content analysis is an effective research method for analysing how different audiences respond to and interpret both scientific topics and science more generally. As a research method it is widely understood, is relatively inexpensive to conduct, and can also be used to examine any written document [7]. It is also a well-established method for conducting science communication research, on topics ranging from the perception of science in the news [8] to investigating the factors that affect science channel popularity on YouTube [9].

So why would you want to use poetry as a data set for conducting a conventional content analysis?

Poetry offers a way for multiple audiences to engage with scientific topics, and to establish their own common language in the process, presenting an opportunity for them to express themselves in a manner that matches their own needs and expertise. In contrast to the scientific jargon that can both alienate non-scientific audiences and act as a barrier to effective science communication, poetry enables the writer to talk about a scientific topic in a manner that is personal to them, helping them to find the words and phrases that are needed to communicate with others more effectively [10].

In short, poetry enables the poet to talk about a topic in their words, their language, and with a context provided by their own lived experiences. Similarly, while several other methods exist for the analysis of poetry (e.g. ethnography, grounded theory), the method that I will present in this Chapter is grounded in conventional content analysis because of its ability to highlight both the context and the content of the chosen text, which for a subjective medium such as poetry is essential.

A traditional approach to analysing data during conventional content analysis [11] would be to begin by identifying 'meaning units' in the text, condensing these down to smaller units, and then labelling these units with codes. These codes would be chosen to describe each meaning unit, after which different codes would be grouped into thematic categories according to content and context. Following this approach, the analysis might then focus on emerging narratives that express an underlying meaning of the text and which could be directly related back to any research question [12]. This process is summarised in Fig. 4.1.

Any approach which utilises a qualitative content analysis should be guided by the following seven steps: formulate research questions, select sample to be analysed, define the codes to be applied, outline the coding process, implement the coding process, determine trustworthiness, and analyse the results of the coding process [4]. These seven steps form the basis of the poetic method that I will now define, which describes how poems might be used as a form of data to provide further insight into the interpretation of scientific topics and how they are communicated.

Fig. 4.1 A traditional approach to conventional content analysis

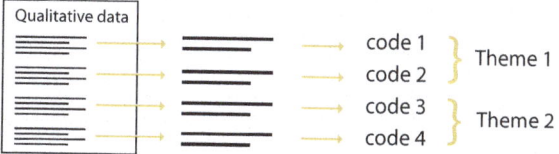

4.3   Poetic Content Analysis

4.3.1   *Formulate Research Questions*

The first step in conducting a poetic content analysis is to define your research questions. There might be one or several of these, but they are essential for underpinning the research that you are about to carry out, and as such they should both align with your own theoretical perspectives and be suitable for the use of poetic content analysis as a method. To do this, begin by asking yourselves two questions: what do I want to find out, and how will analysing poetry enable me to do this?

It might be that the answers to these two questions are intertwined. For example, you might want to investigate how poets have written about both weather and climate over the past 200 years, to consider the trends in which these two terms have been used and the impact that this might have in communicating climate change to different audiences [13]. However, it is still necessary to consider why it is that you want to look at poetry specifically, and what learnings you think you might gather from such an analysis. In this example the research questions (RQ) might be phrased as such:

RQ1   to what extent have trends in the use of the terms 'weather' and 'climate' changed over the past 200 years for poems written in the English language?

RQ2   what might these trends tell us about the way in which climate change is communicated to different audiences?

Similarly, it might be that you have a specific science communication research question in mind but are unsure of how (or why) to use poetic content analysis to answer this. Imagine, for example, that you wanted to investigate vaccine acceptability amongst elderly patients [14]. One of the ways that you might do this is to ask representatives from this audience to write poems about their attitudes and experiences in relation to vaccines and then analyse their responses. As such a suitable research question might be:

RQ   what does poetry reveal about vaccine acceptability amongst elderly patients?

In constructing your research questions it is also a good idea to keep in mind Betteridge's Law of Headlines [15], which stipulates that any headline that ends in a question mark can be answered by the word 'no'. Adapting Betteridge's Law for the purposes of research, we might ask ourselves:

Is it possible to answer this research question with a simple yes/no answer?

If so, then consider rephrasing it. For example,

RQ   does poetry reveal anything about the way in which first-year undergraduate students perceive the importance of ethics in science?

Would be better phrased as:

RQ   to what extent can poetry be used to better understand how first-year undergraduate students perceive the importance of ethics in science?

Crafting your research questions in this manner ensures that they are harmonious with the interpretivist theoretical perspective that underpins this particular research method, which in turn helps to confirm that there is consistency in your wider research methodology.

Exercise 4.1: Formulate a Research Question
Think about a specific science communication topic that you would like to investigate and create one or two research questions for which poetic content analysis would be an appropriate method to adopt. In creating your research questions try to avoid those which could be answered with just a yes/no answer, as they are not particularly fitting with poetic content analysis as a research method.

4.3.2 Select Poetry

Selecting the poetry to analyse will largely be determined by the way in which your research questions frame your study and will tend to fall into one of two categories. Either you will be using poetry that has previously been written about a specific topic, or else you will be soliciting the creation of original poetry. Imagine the following research questions:

RQ1   how have interpretations of 'space travel' changed in poetry over the past century?

RQ2   what do these trends tell us about how poetry might be used to engage different audiences with space travel in the future?

Such research questions would lend themselves to the use of poetry that has already been written, but how would you go about finding and selecting such poetry? The first thing that you need to do is to make a list of justifiable search constraints, each of which must be carefully rationalised to establish the reliability of your method (see Sect. 4.3.6). In the above example you might limit yourself to poems that explicitly use the phrase 'space travel', or which contain the words 'space' and 'travel', just as you might do if you were utilising a more traditional qualitative research method, such as a systematic literature review [16]. You might also try to ensure that you have at least one poem per decade, and that the poetry you select is representative of a diverse range of voices. Similarly you might limit yourself to poetry written in one specific language (in which you are fluent), or else have an author-sanctioned translation of the poem.

In addition to the science poetry resources discussed in Chap. 2, there are a large number of poetry databases that you might use to conduct your search, with the *Poetry Foundation* [17], *The Poetry Society* [18], and *The Poetry Archive* [19] offering three of the most comprehensive. Social media and subject specific mailing lists are also a great resource for locating those poems that are perhaps more obscure or hard to find. For example, with the above research questions you might get in contact with a local astronomer group or a well-known space scientist on Twitter to see if they or their associates know of any poetry for you to include in your selection. I will discuss how to further filter this selection of poetry and how to know if you have selected enough poetry for your study in Sects. 4.3.3 and 4.3.4, respectively.

The other option that you have for selecting your poems is in working with an audience to create them. Take the following research question:

RQ   how can poetry be used to help give voice to those people living with dementia?

For such a research question, you might work with the specific audience in question (in this instance people living with dementia, including their carers) to create poetry which could then be analysed using poetic content analysis. In creating such poetry it is essential that all of the participants (or their guardians) have given their informed consent that the poetry they are creating will be used in this way. Ethical approval should also be sought from your research institute, and in the subsequent analysis of the poetry care must be taken to ensure that any identifiable information in the poetry itself is redacted.

Clear instructions for how participants can extract themselves (and their poetry) from the study should also be provided and participating in the study should be seen as beneficial rather than exploitative, especially when working with potentially vulnerable audiences. There should also be appropriate safeguarding put in place if sensitive information is revealed during the poetry writing process, especially if it indicates that the participant might be at risk. The process of writing and analysing poetry can also potentially cause distress to both the participants and the researcher(s), and steps should be put in place for this to be dealt with appropriately. Chap. 6 will provide a detailed account for how to work ethically with an audience to create poetry that can be used to answer this type of research question.

Your research questions might also lend themselves towards a combination of the above two approaches. For example, you might want to find out what has been written about a specific scientific topic by poets of the past, and then compare your findings to a modern audience's attitudes towards this topic. There is also the option of using your own science poetry, although unless your research question is particularly autoethnographic in nature (e.g. 'How have my attitudes towards diversity in science changed over the past decade?'), I would avoid doing so as it might otherwise affect the trustworthiness of the approach (see Sect. 4.3.6).

Exercise 4.2: Select Some Poetry
Take the research questions that you created in Exercise 4.1 and find some poetry to help you answer them. I recommend reframing your research questions to ensure that they can be answered using poetry that has already been

written, rather than that which requires the creation of original poetry. However, if you do want to pursue this direction, then turn to Chap. 6, which will provide more specific detail for how to work with an audience to create poetry that can be used for such research questions.

If you are selecting a research question that can be answered using existing poetry, then begin by creating a list of search criteria, each of which should be fully justifiable. Use the *Poetry Foundation, The Poetry Society,* and *The Poetry Archive* to start to make a list of poems according to these criteria and reach out to the appropriate communities to see if they can recommend any more that you might otherwise have missed.

4.3.3 Read Poems

Once you have selected your poems, the next thing that you should do is read them, so that you become comfortable with them as a data set. When doing qualitative content analysis with any textual data, it is important that you become familiar with it before you start to assign any codes [20]. Doing so will help to give you confidence in handling the poetry, and it is also an essential step to help you 'clean' the data set before you begin the next stage of the research method.

Data cleaning comprises detecting and removing errors and inconsistencies from these data to improve its quality [21]. This might involve removing spurious data points, negating incomplete sets of survey responses, or even correcting for naming conventions in the files themselves. When dealing with poetry as a data set, an initial reading of the poems can help you to decide which poems to take forward for the subsequent coding, and which are not relevant for your research questions.

If you are selecting poetry that has been purposefully written in response to your research questions, then cleaning the data is relatively straight forward, and will centre on removing those poems for which you either lack the informed consent of the participant to use in this manner, or which contain identifiable information that should be redacted prior to the next stages of investigation. This might be explicit, such as the participant revealing their name or address, or else it might be a phrase or anecdote that is immediately recognisable as being from that participant. Reading the poems at this early stage and redacting any information that could inadvertently be traced back to one of the participants in your study will help to ensure an ethically sound approach to your research method (see Chap. 6 for more details on how to conduct your work ethically).

If you are selecting poetry that has already been written, then cleaning the data will instead take the form of filtering out any poetry that upon reading is not relevant to answering your research questions. By way of an example, suppose you had the following research question:

RQ   how have atmospheric clouds been represented in poetry over the past 300 years, and to what extent do any trends relate to how our scientific understanding of their formation has also changed during this time period?

An initial search for poems that contain the keyword 'cloud' would also likely contain those poems that use the word as a verb to imply something becoming less clear. Similarly, more recent poems might refer to a computer network where files and programs can be stored. These are easily identifiable uses of the word cloud, whose poems can be removed from your selection. However, you might also want to remove poems depending on the way in which the word is used in context, even if the meaning of the word is itself correct. Suppose for example that a poet likens someone's complexion to a cloud. Should this poem be included or not? This very much depends on both your research question, and also the set of justifiable selection criteria with which you choose your poems.

In the case of the above research question, you may also wish to exclude those poems that use clouds as a metaphor or a simile because you are primarily concerned with the ways in which poets observed the physical characteristics of clouds. Alternatively, you may wish to include such poems because you believe that the ways in which poets have changed their usage of clouds as figurative language could also reveal something about the way in which the perception of their physical characterises have likewise changed. Such a choice should be guided by both your research questions and your underlying theoretical perspective.

However you decide to clean your data, be sure to keep a record of those poems that you have removed and your rationale for doing so, as such auditing will help to improve the trustworthiness of your approach. There is a degree of subjectivity that comes through the reading and cleaning of poems as data, the implications of which will be discussed further in Sects. 4.3.6 and 4.4.6.

Exercise 4.3: Read Some Poetry

Read the selection of poems that you identified in Exercise 4.2, alongside the research questions that you developed in Exercise 4.1. Which of the poems can be cleaned from your data set because they are not relevant with respect to your research questions? How are you able to justify this cleaning process so that it could be replicated by others? Make a note of those poems that you have discarded, alongside a clear rationale for why you have decided to negate them.

4.3.4 Assign Codes

Having read and cleaned your poems, the next step is to start assigning codes.

As discussed in Sect. 4.2, a traditional approach to coding data during qualitative content analysis would be to begin by identifying 'meaning units' in the text, condensing these down to smaller units and then labelling these units with codes. However, poetic content analysis utilises a different approach to its treatment of condensed meaning units, negating them in favour of broader labels (a 'code' in this context is just another word for a 'label'). Such an approach is necessary, because in addition to overly short meaning units leading to fragmentation in the coding process (thereby affecting the reliability of the method) [22] poems, unlike more traditional qualitative data sets (e.g. interview transcripts or survey responses), have been crafted by the author so that every line has 'meaning'. As such each line of the poem could already be considered to be a meaning unit and should not be condensed further.

Assigning an overall meaning or tone to the poem as a whole is not necessarily helpful at this stage, unless such an approach is determined by your research question. For example, you may have a research question which encourages a phenomenological approach to be adapted. Such an approach to research is one that seeks to describe the essence of a phenomenon by exploring it from the perspective of those who have experienced it [23], and so the lived experiences of the researcher(s) is considered an essential part of the analysis. For example, a research study that looked at the different ways in which a scientist and a non-scientist interpreted poems written about environmental change might adopt such an approach [24].

But what do I actually mean when I say that you should code or label segments of a poem?

Having read through all of the poems during the reading and cleaning phase of this method you should already be familiar with them. The next step is to read each of them again carefully alongside your research questions, and to label lines of the poem which emerge in response to these questions. The following example shows what this looks like in practice.

Imagine you have the following research question:

RQ-W   how do poets from different countries interpret the relationship between humans and whales?

Having selected, read, and cleaned your poetry, the first poem on the list to begin the coding process is one that is entitled 'Whaling Away', written (by me) about research which found that sperm whales shared behaviours to outmanoeuvre 19th-century human hunters [25]. Figure 4.2 shows how I have labelled this poem in response to the research question, with Table 4.1 summarising the codes in terms of both a description and the frequency of their occurrence. As you can see, it is possible for certain sections of the poem to be assigned multiple codes, and for some sections to be assigned no codes at all.

As each new code is realised you should go back through the poems that have previously been coded to see if these also contain any lines that could also be labelled with this newly emergent code. You should then read all of the poems in full again and make sure that each of them has been coded accurately and that a saturation of emergent codes has been reached, i.e. that there are no new codes to emerge—this

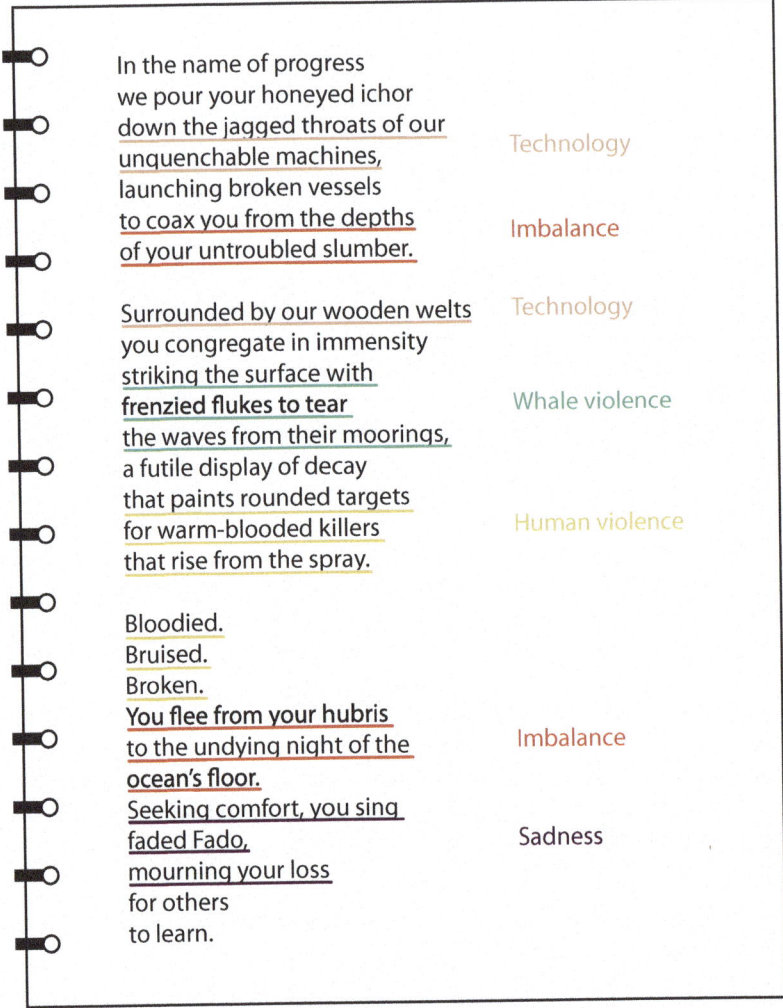

Fig. 4.2 An example of the coding process using the poem 'Whaling Away' in relation to RQ-W

Table 4.1 The codes that emerged from the poetic content analysis of 'Whaling Away', including their descriptors and frequency of occurrence

Code	Description	Frequency
Human violence	Refers to humans being violent	1
Imbalance	Refers to an imbalance in natural behaviours	2
Sadness	Refers to a sad emotion	1
Technology	Refers to a machine or artificial instrument	2
Whale violence	Refers to whales being violent	1

saturation is also an indication that you have collected enough data to answer your research question (see Sect. 4.3.6).

After checking for saturation you should then read through each of the individually coded segments again and make sure that they are appropriate for the code to which they have been assigned. At every stage be sure to document the process and make a note of the number of codes and their frequency of occurrence, and how this changes as you moderate the process. Keeping such a fastidious record helps to ensure that the process is well audited, thereby improving its trustworthiness as a research method (see Sect. 4.3.6).

In terms of 'how' you actually code the poems, you might want to make use of NVivo, a qualitative data analysis computer software package that helps researchers to annotate, label, organise, and analyse qualitative data [26]. However, you can just as easily copy and paste your poems into a word processing document and highlight different segments with different colours (one per code), creating something like that shown in Fig. 4.2. If you are adopting a multiple coder approach (see Sect. 4.3.6) then you should ensure that all of the researchers have access to the same software, and that it can be shared across the different operating systems that are being utilised. Personally, I have found that a combination of Google Docs and Google Sheets is useful for such collaborations, as they are freely available pieces of software that are independent of operating system, and which provide easily accessible version histories to help establish a reliable auditing trail.

Exercise 4.4: Code Some Poetry

Re-read the cleaned selection of poems that you identified in Exercise 4.3 and start to apply codes to sections of the poems, using labels that are relevant to your research questions. Every time a new code emerges from the poetry, go back to the start of your list of poems and see if this code can be applied to those you have already coded. By the time you get to your final poem are any new codes emerging, or have you reached data saturation?

Read through all of the poems again to make sure that they have been coded correctly, and then read through all of the individually coded sections of the poems to make sure that these are also correct. Make a note of the frequency with which each code occurs, and how this might change throughout the process; for example, do you discover that some of your initial codes are actually misaligned with your research questions?

4.3.5 Determine Categories

Once all of the poetry has been coded, and you are confident that data saturation has been reached (i.e. that there are no new codes to emerge from the data), then the next step is to group these codes together into broader categories. These categories should collect together codes that contain overlapping issues with regards to the research questions, to help identify emerging narratives that can be used for the subsequent analysis (see Sect. 4.3.7). There is no set minimum or maximum number for these categories, but as with everything else in this research method, you should be able to justify and fully document the rationale for your grouping of codes.

Continuing with the example of exploring the relationship between humans and whales (RQ-W: how do poets from different countries interpret the relationship between humans and whales?), imagine that having read and coded 50 such poems, the codes shown in Table 4.2 had emerged and data saturation had been reached.

These codes could then be grouped into the categories that are shown in Table 4.3. Note that some categories may only contain one code (e.g. 'Technology'), while others might contain multiple codes but take the name of only one of these (e.g. 'Communication'), or else contain multiple codes but be labelled with the name of a category that is different from any of the codes it contains (e.g. 'Balance').

After this initial grouping of codes has taken place, you should go back through each of the individual coding occurrences (e.g. the 80 occurrences of 'Violence' in the above example) to make sure that these segments of poetry do indeed belong in

Table 4.2 The codes that emerged from the poetic content analysis of a selection of poems (n = 50) in relation to RQ-W

Code	Description	Frequency[a]
Anger	Refers to an angry emotion	10
Communication	Refers to communicating with whales	9
Cruelty	Refers to whaling as being cruel	13
Culture	Refers to a cultural event	12
Extinction	Refers to the extinction of whales	14
Fear	Refers to fear	11
Happiness	Refers to a happy emotion	4
Human violence	Refers to humans being violent	47
Imbalance	Refers to an imbalance in natural behaviours	12
Necessary	Refers to whaling as being necessary	6
Sadness	Refers to a sad emotion	19
Song	Refers to whale song	5
Technology	Refers to a machine or artificial instrument	8
Whale violence	Refers to whales being violent	20

[a]the number of occurrences is not limited to one per poem

Table 4.3 The categories that emerged for RQ-W, alongside their corresponding codes

Category	Corresponding codes	Frequency[a]
Balance	Imbalance, Culture, Extinction, Necessary	48
Communication	Communication, Song	14
Emotion	Sadness, Happiness, Anger, Fear	44
Technology	Technology	8
Violence	Whale Violence, Human Violence, Cruelty	80

[a]the number of occurrences is not limited to one per poem

this category. Reading through all of the segments again will also reveal if there are any other categories to emerge, or if the ones that you have selected are inappropriate with regards to the way in which the codes link with each other and/or to the research questions. These emergent categories are what will be used to form the basis of your analysis, which will be discussed further in Sect. 4.3.7.

Exercise 4.5: Create Some Categories

Go through each of the codes that you created in Exercise 4.4 and start to group them into categories, documenting the rationale for each grouping that you propose. Create a table like that shown in Table 4.3 to document your process, and once you have grouped all of the codes into different categories, read through every individually coded segment of poetry again to make sure that it belongs in the category to which it has now been assigned. Make a note of any changes during this process, and carefully consider the rationale for the links between codes that you are proposing in relation to the research questions.

4.3.6 Confirm Trustworthiness

Before you can begin analysing the emergent categories in response to your research questions, it is necessary for you to first assess the trustworthiness of your research method. The coding and categorisation that I have outlined so far in this method are relatively subjective and is to some degree dependent upon the coder. However, this does not mean that the method is not trustworthy. All qualitative content analysis involves a degree of subjectivity in both the coding and the analysis, as it is simply impossible for any researcher to detach their own lived experiences from the process.

Given this subjectivity however, it is still necessary for the whole research process to be verifiable, to help ensure the integrity of your approach.

The verification of a research method involves checking and confirming to make certain that what you have done is both reliable and valid, thereby establishing the trustworthiness of the process [27]. Reliability refers to how consistently a method measures something, while validity refers to how accurately such a method measures what it is intended to measure. In the case of poetic content analysis, the reliability can be thought of as how likely it would be that the same categories and subsequent analysis emerged if the process was repeated again, while the validity can be thought of as how likely is it that the categories and analysis to emerge from these data are a useful interpretation of the research questions.

The reliability of your approach can be assured by keeping comprehensive notes of the entire process, including a detailed and up to date 'codebook'. A codebook can be thought of as an expanded version of that shown in Tables 4.2 and 4.3, in which you further justify and explore what is meant by each code and category, and why these have been chosen as unique labels. Keeping such a detailed codebook means that you are likely to be able to repeat such an approach yourself, and it also makes clear to other researchers exactly what steps you have taken in arriving at your conclusions.

Similarly, the validity of any poetic content analysis can be established by ascertaining a strong auditing trail throughout the process, providing a thorough justification for the selection and cleaning of the poetry, and ensuring that theoretical data saturation is reached during the coding and categorisation of the poetry; reporting on how, when, and to what degree you have achieved this [27]. In the subsequent analysis, a detailed rationale alongside a careful contextualisation with existing literature (see Sect. 4.3.7) can also help to ensure the validity of the findings.

Given that poetic content analysis is underpinned by an interpretivist theoretical perspective, it is not expected (and is indeed highly unlikely) that any two researchers (or groups of researchers) would create an identical codebook or perform an indistinguishable analysis, even if given the same research questions and an identical set of poems. However, as with all research that is aligned with such a theoretical perspective, it does not make the findings any less valid. In fact it is precisely this difference in interpretation which assures the validity of the approach (providing that the trustworthiness has been established and maintained, as discussed above), as it demonstrates both the strength of the research design and the appropriateness of the method for answering the research questions [28].

Working with multiple coders to perform a poetic content analysis can also further aid the reliability and validity of your approach. If you do adopt a multiple coder approach, then establishing accurate codebooks is essential. I would also recommend for each of the coders to conduct each stage of the poetic content analysis individually, before meeting to discuss their processes at each stage and agreeing upon a set of co-created codes and categories to be used.

For example, if three researchers were performing the coding of the RQ-W, then the first stage (after selecting and cleaning the data) would be for each of the researchers to individually code the poems. The researchers could then meet to agree upon a mutual set of codes, and then individually and collaboratively re-code the poems according to this shared codebook. A similar process could then be repeated for the categorisation of the codes, and also the interpretation of these categories through the subsequent analysis.

Such an approach is not essential for poetic content analysis however, and indeed sole coders can still achieve high trustworthiness providing that the appropriate steps are taken to assure reliability and validity (i.e. good documentation, justification for selection, the creation of clear and detailed codebooks, and evidence of data saturation).

> **Exercise 4.6: Consider Your Trustworthiness**
> Go back through the coding and categorisation that you performed in Exercises 4.4 and 4.5, respectively. Do you have a detailed audit trail that would enable other researchers to follow what you have done at every stage? How understandable is your codebook? If you had to start the process from the beginning, would you arrive at the same set of categories given the documentation that you have provided?

4.3.7 Analyse Results

The final stage in this research method is the analysis of the results. In short this involves taking each of the categories in turn, and then writing about them in relation to the original research questions. This would normally also involve using certain sections of the poems to evidence your argument and contextualising these discussions with relevant findings from existing literature in fields of related research (e.g. peer-reviewed papers, monographs, technical reports). During this process one or several themes may also emerge from the categories, which could be used to answer the research questions directly, although this might not always be the case. In developing a theme, you will be looking for any commonalities and/or overlaps, in a manner analogous to the emergence of the original codes and categories that is described in Sects. 4.3.4 and 4.3.5, respectively.

In analysing your results you should also be cognisant of your own subjectivity as both a researcher and a human being, highlighting this where appropriate. Discussing your own interpretations in this manner and using evidence from the existing research literature to reinforce your findings will help to further establish the validity of both your analysis and the research method as a whole.

To demonstrate what this might look like, imagine that we are conducting the analysis for the categories shown in Table 4.3. In reading all of the sections of poetry that relate to 'Technology', it becomes evident that technology acts as an obstacle against the harmonious co-existence between humans and whales that appears to be independent of both country and historic period. In making this point I might choose to quote several sections of the poetry that evidence this (e.g. 'In the name of progress // we pour your honeyed ichor // down the jagged throats of our // unquenchable machines') and also contextualise this finding with other similar observations that have previously been made by other researchers; for example that the pursuit of new technologies during the industrial revolution were to some extent dependent upon the availability of whale oil [29].

Providing evidence for your own analysis (in the form of relevant sections of poetry) alongside independent research that has made a comparable observation helps to verify the research method and improve its trustworthiness.

Exercise 4.7: Analyse Your Results

Take each of the categories that you established in Exercise 4.5 and use them to provide an analysis of your research questions. For each category make sure that you can provide evidence from the poems that you have coded and try to support these findings with comparable observations in the existing research literature. It might be that there are no such links because you have made a truly novel observation. In this instance, concentrate on finding some alternative evidence from the literature that will help to further contextualise and validate your analysis.

Figure 4.3 presents a visual representation of the poetic content analysis research method, which I will now expand on further in Sect. 4.4 by providing a worked example of what this looks like for each of the seven stages.

4.4 In Practice: A Change of Climate

To further illustrate how poetic content analysis works as a research method, I now present a worked example for each of the seven steps, demonstrating how to move from the formulation of a research question, through to the selection, coding, and analysis of poetry.

4.4.1 Formulate Research Questions

To fully address the global interdisciplinary problem that is mitigating against, and adapting to, anthropogenic climate change, scientists will need the help of other publics. By neglecting to involve non-scientists in this discussion, any such 'solution'

Fig. 4.3 The seven steps of
the poetic content analysis
research method

will fail to account for the expertise of these so called 'non-experts'. While these
publics might not be employed in either scientific research or policymaking, they still
have a large amount of tacit and lived expertise that is potentially vital for creating
and implementing effective mitigation and adaptation strategies [30].

Starting multidirectional conversations between these publics is necessary,
because doing so grants agency, builds trust, and creates a platform for diverse solutions. However, initiating such dialogue can be difficult, especially amongst those
audiences that have not previously engaged in any conversations around climate
change [31]. Furthermore, the way in which climate change should be communicated
is also debated, with several communication experts convinced that identifying the
beliefs and needs of the individual is key, while others insist on a facts-driven method
of delivery [30].

Poetry offers a potential medium through which to initiate these conversations, by
giving voice to people other than scientists in relation to environmental change and
the climate crisis [24]. By examining what poems have to say about current climate
change mitigation and adaptation efforts, we might also be offered an insight into the
challenges that we face in enabling society to take positive action. This hypothesis
is represented by the following research question:

RQ-A What do poems written about climate change reveal about current attitudes
       towards climate change mitigation and adaptation efforts?

4.4.2 Select Poetry

In the summer of 2017, my colleague Dan Simpson and I initiated a global poetry competition. A challenge to find 20 poems that spoke about climate change in different voices, and from beyond an exclusively scientific perspective. The only stipulations that we set were that the poems had to be about climate change (however the poet might perceive this) and that they had to be 40 lines or fewer (this was to aid with the typesetting of the subsequent collection).

We received 174 entries from 23 countries in five different languages, and after a rigorous selection process 20 poems were chosen and published as the collection *A Change of Climate* [32]. The publication of this book was made possible by funding from the Greenhouse gAs Uk and Global Emissions (GAUGE) project [33], with all of the profits going to support the work of the *Environmental Justice Foundation*, an organisation that works to protect both our people and our planet, by campaigning for human rights and the rights of the environment.

The 20 poems from *A Change of Climate* will be used to answer this research question. As RQ-A was formulated after the production of *A Change of Climate*, this is an example of using poetry that has previously been written about a specific topic, rather than the use of original poetry that has been specifically created in response to a research question. The reason that these poems have been selected for this analysis is because they provide a diverse collection of voices that interpret climate change from beyond a purely scientific perspective.

4.4.3 Read Poems

The next step in this research process is to read the 20 poems, specifically with RQ-A in mind. I present three of the poems, as they appear in *A Change of Climate*, with the kind permission of the authors. I have chosen these three poems to illustrate the cleaning of the data that takes place in this stage of the research method.

We are no longer interested in the sea *by Michael Conley*

We are no longer interested in the sea.
The sea is a tiresome old man shouting.
down dementia's cushioned corridors.
and we are not fooled by its bluster. In fact.
we are sick of the sea. We are sick.
of its desire to relive former glories,
its gatecrashery of our parties,
its pathetic attempts at coup d'état.
We should kill it. We should go.
to the beach at night and pelt the sea.

with stones until our breath pricks.
our lungs like a swallowed anemone.
Bring everything you have: kettles,
refrigerators, old washing machines.
Pitch them in. Force everything we have ever built.
down its open gullet and lose yourselves.
in the ecstasy of it. We will drown the sea.
in solidity and we will walk upon it.
in our thousands. The earth and the sky.
will be forced to sit up and take notice.

SLAG *by Emily Cotterill*
I have loved coal,
like a teenage girl loves an older guitarist.
with a rough black smudge of eyeliner.
I have built my life on it,
screamed down decades for it,
COAL NOT DOLE –
bared my whole soul for it.
but old women gossip about the pit,
I know the world has had enough of it.
Coal – with its head full of history,
strong arms, filthy engines, heavy,
the small town sex of it.
Broken bodies, white knuckle wives,
the silence of canaries – has risen.
from slag heaps and pit heads.
to thick air spluttering into anyone born.
late with withered old miners' lungs.
I have loved coal but recently,
when I sit in the fresh place built.
on the scar of my grandfather's pit.
I have loved birdsong, greenspace,
the safety and hope of it –
wind turbines, rising white beacons,
sharp armed, slicing clean paths.
to a future.

*This poem also appears under the title of 'I Have Loved Coal' in *The Day of the Flying Ants* [34].

Karner Blue *by Carrie Etter*

'... a place called Karner, where in some pine barrens, on lupines, a little blue butterfly I have described and named ought to be out.'—Vladimir Nabokov

Because it used to be more populous in Illinois.

Because its wingspan is an inch.

Because it requires blue lupine.

Because to become blue, it has to ingest the leaves of a blue plant.

Because its scientific name, *Lycaeides melissa samuelis*, is mellifluous.

Because the female is not only blue but blue and orange and silver and black.

Because its beauty galvanizes collectors.

Because Nabokov named it.

Because its collection is criminal.

Because it lives in black oak savannahs and pine barrens.

Because it once produced landlocked seas.

Because it has declined ninety per cent in fifteen years.

Because it is.

In my initial reading of the 20 poems, I noted that 16 of them referred to climate change adaptation or mitigation efforts. Some of these were positive (such as the transition towards green energy discussed in 'SLAG'), while others were negative (such as the futility of action displayed in 'We are no longer interested in the sea'. Four of the poems (such as 'Karner Blue') made no clear reference to climate change adaptation or mitigation efforts or attitudes, and as such they were 'cleaned' and removed from the next stages of the research method.

There is of course a degree of subjectivity in this cleaning process, and some other coders might have excluded 'We are no longer interested in the sea' from the selection process as it does not make explicit reference to mitigation or adaptation efforts (which is the case for 'SLAG'). However, I would err towards leniency in the cleaning process providing that a clear justification can be made for each poem's inclusion. In this instance I have decided to include 'We are no longer interested in the sea' because it talks about actions that humans are taking to affect the environment and/or climate, which I think justifies its inclusion for the subsequent analysis. Similarly, I have excluded 'Karner Blue' from the subsequent coding process because it is concerned primarily with an endangered subspecies of small blue butterfly and does not refer (explicitly or otherwise) to climate change mitigation or adaptation efforts from a human perspective.

4.4.4 Assign Codes

After reading the 16 poems that referred to climate change adaptation or mitigation efforts, I started the coding process. In creating these codes, I began by reading the poems in the order in which they appeared in *A Change of Climate*, and every time a new code emerged, I went back to the start of the 16 poems and looked to see if there were any further examples of this code that I might have missed. Through this process I assigned the codes that are described in Table 4.4. To help me in this process, and to further establish the trustworthiness of my approach, I also made a note of some specific examples for each code. Doing so is also useful for the ensuing

Table 4.4 The codes that emerged from the poetic content analysis of a selection of poems (n = 17) in relation to the RQ-A

Code	Description	Example	Frequency[a]
Abandon	Discusses the idea that we should leave Earth and start elsewhere	'Earth is a forlorn shore. // Soon we touch the sky' From '21__'	2
Belated	Explores the fact that actions have come too late to make a difference	'we // incredulous and confident // understood it was // too late' From 'Îles du Vent'	5
Cyclical	Explores the concept that Earth/the climate will somehow 'right itself' independently of human action	'And Chaos scythes it clean, comforting, // *It'll soon grow back again.*' From 'Gaia goes to the Hairdresser'	4
Despair	Demonstrates a sense of dread that leads to inaction	'There are holes in the sky!' From 'Needlework'	2
Future	Displays hope for a future in which some aspects of climate change have been averted	'wind turbines, rising white // beacons, // sharp armed, slicing clean paths // to a future.' From 'SLAG'	3
Joy	Shows pride or glee in the damage that is being caused through either action or inaction	'We want to die clean.' From 'Swimming Lesson'	3
Neglect	Evidence of purposefully avoiding adaptation or mitigation efforts	'We should kill it. We should go // to the beach at night and pelt the sea // with stones' From 'We are no longer interested in the sea'	6
Pointless	Demonstrates an inevitability, despite any action that might be taken	'We're too small and single in our homes' From 'The Honey and the Polar Bear'	9
Responsibility	Evidence that humans have a responsibility to nature	'we are entangled and entwined // one // only you hold the power to save us' From 'Earth Plea'	5

[a]The number of occurrences is not limited to one per poem

categorisation and analysis, as it helps to highlight examples in the poetry that can be used to support any emergent narratives.

By the time I reached the fifteenth of the 16 poems, there were no newly emergent codes, and so I was happy that data saturation had been reached with respect to RQ-A. I then re-read the four poems that I had previously excluded during the data cleaning stage of the method and decided that one of the poems ('Eschaton' by Andrew Paul Wood) should actually be included in the selection, because two of the emergent codes ('Neglect' and 'Despair') could in fact be applied to sections of the poem in relation to RQ-A. However, including this poem and re-reading it with RQ-A in mind did not reveal any further codes.

In this instance I chose to re-read the four poems that I had excluded because it was a small number, whereas for other research studies this might not always be possible. Table 4.4 reflects this addition (n = 17), and by recording my rationale for doing so, and re-checking to ensure that there were no additional codes to emerge as a result, I have further established the trustworthiness of this approach.

4.4.5 Determine Categories

Table 4.5 presents the four broad categories that group together the codes presented in Table 4.4. In creating these categories I looked for links between the codes that could be used to tie them together (with respect to RQ-A), working from a comprehensive definition that allowed for such a grouping to take place. The definitions (which are in relation to attitudes towards current climate change and mitigation strategies), alongside the corresponding codes are also presented in Table 4.5 to help further document the process.

Following this categorisation, I then went back through each of the individually coded elements for each of the 17 poems and made sure that the assigned category was also relevant in each instance. During this process I did not find any spurious coding or mis-categorisation, nor did I observe any newly emergent codes or categories. This final check gave me further confidence that data saturation had been reached,

Table 4.5 The categories that emerged for RQ-A, alongside their corresponding codes

Category	Corresponding codes	Definition	Frequency[a]
Belated	Belated, Cyclical, Despair	Efforts are already too late	11
Avoided	Pointless, Abandon	Efforts are being avoided	11
Masochistic	Neglect, Joy	There is a willingness for efforts not to succeed	9
Hopeful	Future, Responsibility	Efforts suggest a successful route for action	8

[a]The number of occurrences is not limited to one per poem

and that the four categories to have emerged were representative of the collective narrative of the poems in response to RQ-A.

4.4.6 Confirm Trustworthiness

As discussed in Sect. 4.3.6, the validity and reliability (and hence the trustworthiness) of the research method can be established by ascertaining a strong auditing trail throughout the process. This includes a justification for the selection and cleaning of the poetry, ensuring that theoretical data saturation is reached during all stages of coding and categorising the poems, and being open and transparent in the creation of any codebooks. As can be seen from Tables 4.4 and 4.5 and the surrounding discussions, there is a clear documentation for all of the steps that I have taken during the poetic content analysis in relation to RQ-A. This detailed auditing means that other researchers could follow the steps that I have taken in arriving at the four emergent categories shown in Table 4.5. The interpretivist nature of this research method means that other researchers may arrive at different categories, but that does not invalidate the trustworthiness of the approach.

As will be evidenced in Sect. 4.4.7, the analysis of the results and the contextualisation of the four emergent categories with additional research literature, alongside evidential segments of the coded poetry, help to give further confidence that the way in which I have answered RQ-A is a useful (and thus valid) interpretation of the research question.

4.4.7 Analyse Results

I will now describe in detail my analysis for each of the four categories to emerge from these poems in response to RQ-A.

Belated

These are those poems that consider current climate change mitigation and adaptation efforts to be ineffectual, because it is already too late to reverse the negative effects of anthropogenic climate change. This opinion is apparent in the following lines from Camille Banks' 'Leeward Islands':

> and when the wind hit
> with its name of clairvoyant
> we
> incredulous and confident
> understood it was
> too late

Another example of this opinion can be found in these lines, from the poem 'Eschaton' by Andrew Paul Wood:

Long since
all towns became Venice or
shut their doors, desiccated… And choked on dust.

What is the point in acting we are asked, if the materials and systems that we find ourselves interreacting with on a daily basis resist such efforts? This attitude of it being 'too late' to mitigate or adapt to the effects of climate change is extremely damaging towards any efforts to persuade individuals to take positive action. Such attitudes are further entrenched by attempts to 'scare' people into action; instead it is necessary to demonstrate to individuals that it is not too late, and to promote positive examples of how they might effectively engage with mitigation efforts [35]. In addressing this feeling of helplessness it is also important to highlight how any potential solutions should not be the sole responsibility of the individual, but that instead government and industry both have a responsibility to adopt sustainable practices beyond token gestures [36].

Avoided

Poems that fell into this category express how it is often easier for the individual to continue as usual, and to pretend that nothing has happened, rather than to focus on action through mitigation and adaptation. Such an attitude is evident in the following lines from 'The Dead Zone Arranged by People' by Alla-Valeria Mikhalevich:

the environment is almost gone -
better even not to look out the window.

Another example that expresses this opinion are the following lines taken from '21__' by Amlanjyoti Goswami:

We knew this for a hundred years
Yet nobody did anything.

Humans tend to assess risks based on personal experience, and the relatively slow and gradual changes that are brought about by climate change (say in comparison to the risks of being hit by a car when crossing the road) make it difficult for many individuals to assess it at an individual level [37]; as such they are less likely to consider it a serious threat. This attitude of willingly looking away is also likely informed by the notion that 'someone else will figure it out' or that 'science has the answers'. However, despite advances in climate technologies such as geoengineering and carbon capture, the effects of climate change cannot be solved through research alone, i.e. there is no Manhattan Project to tackle climate change [38].

Masochistic

These poems are those that explore the notion that climate change mitigation and adaptation efforts will not be successful because humans are willingly bringing about their own destruction. This is exemplified throughout Michael Conley's 'We are no longer interested in the sea', but is especially evident in the following lines:

> Bring everything you have: kettles,
> refrigerators, old washing machines.
> Pitch them in. Force everything we have ever built
> down its open gullet and lose yourselves

The following lines from 'Swimming Lesson' by Julian Dobson express a similar opinion:

> We're filling our faces with containers of bleach, washing up liquid,
> hair products, makeup remover. We want to die clean.

The poems that are categorised within this narrative might be seen as a spurious dataset, as poets and their work (perhaps unfairly) are often considered to be more masochistic when compared to other publics [see e.g. 39]. However, these poems reveal an attitude that is averse to enacting positive change, i.e. the need for humans to dominate and subjugate their surroundings. Despite the sustenance that is provided to us by our planet, a large part of humankind's recent history has been one of conquest, a desire to overcome nature and to occupy wild lands and remote areas [40].

There is at times a rather extreme 'us vs. them' attitude that is adopted in our thinking with regards to nature (especially in Western thought), and in reminding us of this, these poems seek to highlight a further obstacle that must be overcome in our efforts to mitigate and adapt to climate change: we must be willing to work in symbiosis with our planet rather than to act in a manner that asserts our dominance.

Hopeful

Poems that belong in this category are those that express hopefulness towards the success of current climate change mitigation and adaptation strategies. Such hopefulness is evidenced in these lines from Emily Cotterill's 'SLAG' in relation to green energy solutions:

> wind turbines, rising white beacons,
> sharp armed, slicing clean paths
> to a future.

Another example of this hopefulness is evident in Kim Goldberg's 'The Keys of the Piano':

> And the people blinked
> but recovered.

Which explores the idea that it is not in fact too late, and that that the current climate catastrophe could actually serve as a wake-up call to halt future environmental degradation [41], standing in stark contrast to the belated apathy to emerge from the analysis of the other three broadly negative categories.

Similarly, the following segment from Marjorie Moorhead's 'Wandering the Anthropocene' explores the idea that the negative impacts of climate change might force collective action and help to re-establish community amongst various other acts of tribalism [42]:

Striving for existence reclaimed

Community found in terms un-dictated

These poems highlight that there are certain things to be hopeful about with regards to current climate change mitigation and adaptation efforts, and without whitewashing the situation they demonstrate that the way in which such efforts are framed potentially changes the attitudes of response in climate change communications [43] and how this in turn can influence the intentions of various publics to do something about it [44, 45].

In drawing this analysis to a conclusion, let us re-visit RQ-A, which asked: what do poems written about climate change reveal about current attitudes towards climate change mitigation and adaptation efforts?

From the poetic content analysis that has been performed in this example, there is evidence of an emergent theme: that these poems (and their poets) provide not only a translation of attitudes towards climate change mitigation and adaptation efforts, but also an interpretation of our planet's needs, highlighting the challenges to be overcome to better engender individual and collective action. These poems make clear the need for scientists and communication experts to provide positive visions that can help people to take ownership of the current situation; visions that can convince them that it is not too late, which demonstrate to them that there is an individual risk if we simply choose to look away, and which highlight the opportunity for a more symbiotic relationship with our planet (and the hope that this is achievable).

Despite the relatively small number of poems that have been analysed here, there is also evidence that they provide both a potential framework and a translation for the development and delivery of effective climate change mitigation and adaptation strategies, and as such a starting point for the conversations that need to take place between our planet's various publics.

4.5   Summary

This Chapter has presented an outline of poetic content analysis, a research method for developing and interrogating science communication research questions by using poetry as data. As well as introducing this method and how to use it in practice, this Chapter has also discussed the importance of considering your theoretical perspectives, and the need to ensure that there is a good alignment between research question and research method.

Having read this Chapter, you should now be in a strong position to conduct your own research using poetic content analysis. Working your way through all of this Chapter's exercises will help you to become familiar with this research method and will also present you with the grounding to use poetic content analysis as a tool to develop and conduct your own interdisciplinary science communication research using this approach. In Chap. 5, I will go on to discuss an alternative research method, which rather than using poetry as data, uses poetry as a tool through which to analyse more traditional qualitative data sets.

4.6 Suggested Reading

For an accessible description of theoretical perspectives, ontologies, epistemologies, and methods, Donna Berryman has written an excellent article which uses cooking as a metaphor for better understanding such terminology and its application to academic research [46]. Further examples of poetic content analysis can also be found in the following two studies: "'This bookmark gauges the depths of the human": how poetry can help to personalise climate change' [47] and 'In my remembered country: what poetry tells us about the changing perceptions of volcanoes between the nineteenth and twenty-first centuries' [48]. The former of which investigates the capability of poets to interpret climate change for non-scientific audiences utilising a sole coder, while the latter makes use of two coders to investigate the relationship between society and volcanoes over the past three hundred years. Both of these studies can be read alongside Sect. 4.4 to give further evidence for how poetic content analysis works in practice.

4.7 Further Study

The further study section in this Chapter is designed to help you practice using the poetic content analysis research method, and to consider how you might use it to develop and interrogate your own science communication research questions.

1. **Find a coding buddy.** Using the research that you have developed in the other exercises in this Chapter, find a colleague who would be interested in conducting this research with you, and begin the coding and categorisation process again. At each stage make sure that you have agreed upon a process and a codebook and keep detailed notes of every decision that you make (both individually and as a collective). How does this process and the subsequent analysis differ from your findings as a sole coder? To what extent has the validity and reliability of your findings been affected?
2. **Formulate a new question**. Go back to Exercise 4.1 and formulate a research question for a new research study. Work your way through Exercises 4.2–4.7 to select poems, read them, code and categorise them, examine the trustworthiness of your approach and analyse your findings. Was anything different in the way

that you conducted your research this time around now that you are more familiar with the method?

3. **Consider your approach**. As discussed in Sect. 4.1, poetic content analysis is underpinned by an interpretivist theoretical perspective. However, does this align with your own views and attitudes towards how (and why) research can be conducted? If not, is there a way to adopt or adapt this method so that it is more congruent with your own underlying approach to how science should be both researched and communicated?

References

1. Scotland J (2012) Exploring the philosophical underpinnings of research: relating ontology and epistemology to the methodology and methods of the scientific, interpretive, and critical research paradigms. Engl Lang Teach 5(9):9–16. https://doi.org/10.5539/elt.v5n9p9
2. Gray DE (2013) Doing research in the real world. SAGE Publications Ltd., London
3. Joffe H (2012) Thematic analysis. In: Harper D, Thompson A (eds) qualitative research methods in mental health and psychotherapy: a guide for students and practitioners. Wiley-Blackwell, Chichester
4. Hsieh H-F, Shannon SE (2005) Three approaches to qualitative content analysis. Qual Health Res 5(9):1277–1288. https://doi.org/10.1177/1049732305276687
5. McDermott JF Jr, Porter D (1989) The efficacy of poetry therapy: a computerized content analysis of the death poetry of Emily Dickinson. Psychiatry 52(4):462–468. https://doi.org/10.1080/00332747.1989.11024470
6. Hoover DL, Culpeper J, O'Halloran K (2014) Digital literary studies: Corpus approaches to poetry, prose, and drama, vol 16. Routledge, New York
7. Vitouladiti O (2014) Content analysis as a research tool for marketing, management and development strategies in tourism. Procedia Econ Finance 9:278–287. https://doi.org/10.1016/S2212-5671(14)00029-X
8. Shea NA (2015) Examining the nexus of science communication and science education: a content analysis of genetics news articles. J Res Sci Teach 52(3):397–409. https://doi.org/10.1002/tea.21193
9. Welbourne DJ, Grant WJ (2016) Science communication on YouTube: factors that affect channel and video popularity. Public Underst Sci 25(6):706–718. https://doi.org/10.1177/0963662515572068
10. Bullock OM, Amill DC, Shulman HC et al (2019) Jargon as a barrier to effective science communication: evidence from metacognition. Public Underst Sci 28(7):845–853. https://doi.org/10.1177/0963662519865687
11. Braun V, Clarke V (2006) Using thematic analysis in psychology. Qual Res Psychol 3(2):77–101. https://doi.org/10.1191/1478088706QP063OA
12. Erlingsson C, Brysiewicz, (2017) A hands-on guide to doing content analysis. African J Emerg Med 7(3):93–99. https://doi.org/10.1016/j.afjem.2017.08.001
13. Gowda MVR, Fox JC, Magelky RD (1997) Students' understanding of climate change: insights for scientists and educators. Bull Am Meteor Soc 78(10):2232–2240. https://doi.org/10.1175/1520-0477-78.10.2232
14. Roh NK, Park YM, Kang H et al (2015) Awareness, knowledge, and vaccine acceptability of herpes zoster in Korea: a multicenter survey of 607 patients. Ann Dermatol 27(5):531–538. https://doi.org/10.5021/ad.2015.27.5.531
15. Cook JM, Plourde D (2016) Do scholars follow Betteridge's Law? the use of questions in journal article titles. Scientometrics 108:1119–1128. https://doi.org/10.1007/s11192-016-2030-2

16. Boell SK, Cecez-Kecmanovic D (2015) On being 'systematic' in literature reviews. In: Willcocks LP, Sauer C, Lacity MC (eds) Formulating research methods for information systems. Palgrave Macmillan, London

17. Poetry Foundation (2021) Poetry foundation. https://www.poetryfoundation.org. Accessed 10 Dec 2021

18. The Poetry Society (2021) The poetry society. https://poetrysociety.org.uk. Accessed 10 Dec 2021

19. The Poetry Archive (2021) The poetry archive. https://poetryarchive.org. Accessed 10 Dec 2021

20. Dey I (1993) Qualitative data analysis: a user friendly guide for social scientists. Routledge, London

21. Han J, Kamber M, Pei J (2012) Data mining: concepts and techniques. Elsevier, New York

22. Graneheim UH, Lundman B (2004) Qualitative content analysis in nursing research: concepts, procedures and measures to achieve trustworthiness. Nurse Educ Today 24(2):105–112. https://doi.org/10.1016/j.nedt.2003.10.001

23. Teherani A, Martimianakis T, Stenfors-Hayes T et al (2015) Choosing a qualitative research approach. J Grad Med Educ 7(4):669–670. https://doi.org/10.4300/JGME-D-15-00414.1

24. Illingworth S, Jack K (2018) Rhyme and reason-using poetry to talk to underserved audiences about environmental change. Clim Risk Manag 19:120–129. https://doi.org/10.1016/j.crm.2018.01.001

25. Whitehead H, Smith TD, Rendell L (2021) Adaptation of sperm whales to open-boat whalers: rapid social learning on a large scale? Biol Let 17(3):20210030. https://doi.org/10.1098/rsbl.2021.0030

26. Edwards-Jones A (2014) Qualitative data analysis with NVIVO. J Educ Teach 40(2):193–195. https://doi.org/10.1080/02607476.2013.866724

27. Morse JM, Barrett M, Mayan M et al (2002) Verification strategies for establishing reliability and validity in qualitative research. Int J Qual Methods 1(2):13–22. https://doi.org/10.1177/160940690200100202

28. Cypress BS (2017) Rigor or reliability and validity in qualitative research: Perspectives, strategies, reconceptualization, and recommendations. Dimens Crit Care Nurs 36(4):253–263. https://doi.org/10.1097/DCC.0000000000000253

29. Neves K (2010) Cashing in on cetourism: a critical ecological engagement with dominant E-NGO discourses on whaling, cetacean conservation, and whale watching. Antipode 42(3):719–741. https://doi.org/10.1111/j.1467-8330.2010.00770.x

30. McLoughlin N, Corner A, Capstick S et al (2018) Climate communication in practice: how are we engaging the UK public on climate change? Climate Outreach, Oxford

31. Whitmarsh L, Corner A (2017) Tools for a new climate conversation: a mixed-methods study of language for public engagement across the political spectrum. Glob Environ Change 42:122–135. https://doi.org/10.1016/j.gloenvcha.2016.12.008

32. Illingworth S, Simpson D (eds) (2017) A change of climate. CreateSpace, Scotts Valley

33. Palmer PI, O'Doherty S, Allen G et al (2018) A measurement-based verification framework for UK greenhouse gas emissions: an overview of the Greenhouse gAs Uk and Global Emissions (GAUGE) project. Atmos Chem Phys 18(16):11753–11777. https://doi.org/10.5194/acp-18-11753-2018

34. Cotterill E (2019) The day of the flying ants. Smith|Doorstop Books, Sheffield

35. O'Neill S, Nicholson-Cole S (2009) "Fear Won't Do It" promoting positive engagement with climate change through visual and iconic representations. Sci Commun 30(3):355–379. https://doi.org/10.1177/1075547008329201

36. Ferreira J (2018) Fostering sustainable behaviour in retail: looking beyond the coffee cup. Soc Bus 8(1):21–28. https://doi.org/10.1362/204440818X15208755029519

37. Weber EU (2010) What shapes perceptions of climate change? WIREs Clim Change 1:332–342. https://doi.org/10.1002/wcc.41

38. Yang C-J, Oppenheimer M (2007) A "Manhattan Project" for climate change? Clim Change 80:199–204. https://doi.org/10.1007/s10584-006-9202-7

39. Jones E (2009) In conversation with norm sibum. PN Rev 36(1):46–49
40. Hughes JD (2009) An environmental history of the world: humankind's changing role in the community of life. Routledge, London
41. Mcmullen CP, Jabbour J (eds) (2009) Climate change science compendium. United Nations Environment Programme, Nairobi
42. Shah NY, Wirkus L, Swatuk LA (2018) Can climate change challenges unite a divided Jordan River Basin? In: Swatuk LA, Wirkus L (eds) Water, climate change and the boomerang effect. Routledge, Abingdon
43. Morton TA, Rabinovich A, Marshall D et al (2011) The future that may (or may not) come: how framing changes responses to uncertainty in climate change communications. Glob Environ Chang 21(1):103–109. https://doi.org/10.1016/j.gloenvcha.2010.09.013
44. Dickinson JL, Crain R, Yalowitz S et al (2013) How framing climate change influences citizen scientists' intentions to do something about it. J Environ Educ 44(3):145–158. https://doi.org/10.1080/00958964.2012.742032
45. Bilandzic H, Kalch A, Soentgen J (2017) Effects of goal framing and emotions on perceived threat and willingness to sacrifice for climate change. Sci Commun 39(4):466–491. https://doi.org/10.1177/1075547017718553
46. Berryman DR (2019) Ontology, epistemology, methodology, and methods: information for librarian researchers. Med Ref Serv Q 38(3):271–279. https://doi.org/10.1080/02763869.2019.1623614
47. Illingworth S (2020) This bookmark gauges the depths of the human: how poetry can help to personalise climate change. Geosci Commun 3(1):35–47. https://doi.org/10.5194/gc-3-35-2020
48. Soldati A, Illingworth S (2020) In my remembered country: what poetry tells us about the changing perceptions of volcanoes between the nineteenth and twenty-first centuries. Geosci Commun 3(1):73–87. https://doi.org/10.5194/gc-3-73-2020

Chapter 5
Poetic Transcription

5.1 Introduction

In Chap. 4, I introduced the concept of poetic content analysis and how this can be applied to science communication research. Poetic content analysis is one of several methods of poetic inquiry that can be used by qualitative researchers [1], with poetic transcription being an alternative method that uses poetry in a different manner to that described thus far. Whereas poetic content analysis is concerned with analysing poetry to find emergent narratives in response to a research question, poetic transcription centres on constructing poetry from pre-existing qualitative data sets (e.g. survey responses, interviews, field notes), to create a poem that itself presents an interpretive narrative of the data.

Poetic transcription can be traced back to the concept of 'found poetry', a process which takes existing texts and re-mixes their content to create original pieces of poetry. Found poems can be thought of as a literary collage, created from any source that contains text, from newspaper articles and policy manifestos to letters and graffiti; they can even use other poems as part of their source material. There are many different types of found poetry, with some of the most well-known including: erasure (in which poets take an existing text and erase or redact several sections leaving behind only those elements of the text that they want to highlight), cento (a poetic form composed entirely of lines from poems by other poets), and cut-up (where poets physically cut up or tear fragments of text and re-arrange them together).

Found poetry also exists as a type of qualitative research method; for example as a literature review [2] or autoethnography [3]. However, these methods are themselves pre-dated by poetic transcription, which was first presented as an experimental form of writing by the qualitative researcher Corrine Glesne in her seminal study 'That Rare Feeling: Re-presenting Research Through Poetic Transcription' [4]. In this study, Glesne (herself inspired by the work of Laurel Richardson [5]) defines poetic transcription as the creation of poem-like compositions from the words of interviewees, using interviews conducted with Dona Juana, an elderly Puerto Rican researcher and educator, to demonstrate her process. Several other researchers have since gone on to adopt and adapt this approach, perhaps most notably Monica Prendergast and Sandra L. Faulkner and their work on the use of poetry in social science qualitative research practices [6] and poetic inquiry as feminist methodology [7], respectively.

The method that I present in this Chapter builds on the work of these researchers, and others, to outline an approach for using poetic transcription as a research method through which to interrogate scientific discourse, and to potentially give voice to those audiences who have been underrepresented and underserved by both science and science communication [8].

As with poetic content analysis, the poetic transcription research method that I present here is grounded in an interpretivist theoretical perspective, as it is highly unlikely that any two researchers, having been given an identical data set to interpret, would arrive at the same poem. However, this is not to say that this method is untrustworthy, and there are several steps that should be taken to ensure the reliability and validity of both the final poem and also the research process itself. These will be discussed more thoroughly in Sect. 5.2, but one of the key aspects is in ensuring that the voice(s) of both the texts and the people that created them are well represented.

5.2 A Method for Poetic Transcription

Before I go on to describe this method in detail, I first want to outline the core principles around which it is based. There are many different ways in which to perform poetic transcriptions of qualitative data sets, and I am not saying that my method is 'better' or 'worse' than any of the others. Rather, by presenting these core principles (shown in Fig. 5.1) I hope to demonstrate the rationale and evolution of this method as a form of qualitative analysis.

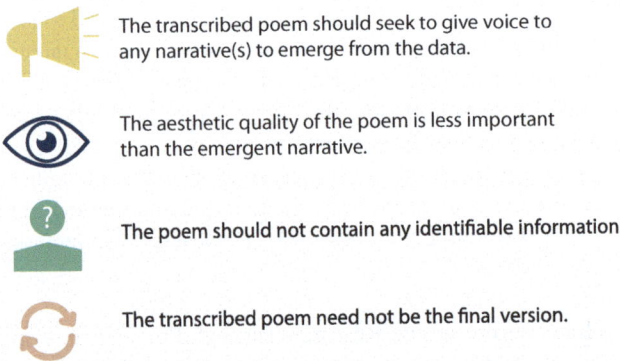

The transcribed poem should seek to give voice to any narrative(s) to emerge from the data.

The aesthetic quality of the poem is less important than the emergent narrative.

The poem should not contain any identifiable information.

The transcribed poem need not be the final version.

Fig. 5.1 The four core principles of poetic transcription

Give Voice

The first of these core principles helps to ensure that any poem is truly representative of both the texts and authors that were used for the transcription. As will be discussed in Sect. 5.2.4, it is first necessary to fully consider the emergent narratives from the data set before any poem starts to fully take shape. This principle also means that the poem that is created during the research process should not contain any additional words or phrases that have been added by the researcher (see Sect. 5.2.5), and that care should be taken to ensure that the words that are used are not presented out of context.

Don't be Led by Aesthetics

The second core principle reminds the researcher of the purpose of the transcribed poem. The poem should exist to help give voice to the emergent narratives, rather than show off the poetic skills of the transcriber. This core principle is essential in the writing stage of the research process (see Sect. 5.2.5), as it might be that the researcher has a choice between an aesthetically pleasing turn of phrase that is incongruous with the emergent narratives and a line that sounds somewhat clichéd, but which is narratively appropriate. In such an instance the researcher should always opt for the latter. This is not to say that the transcribed poems cannot be pleasing to read (and hear), but rather that artistic license should not take preference over the voices that the poem is trying to represent.

Retain Anonymity

The third of these core principles is concerned with the ethical considerations to arise from such a poem. It might be that the data you identify for use in your poetic transcription (see Sect. 5.2.2) has not been redacted and contains information that might identify the person who created the original text (e.g. interviewee, survey respondent). Such identifying information may either by explicit (e.g. name, address, job title), or implicit (e.g. phrase, anecdote, personal history). It is important for any such identifiable information to not be included in the transcribed poem, especially if data are being used where people might not have given their explicit consent for it to be transcribed in this way (see Sect. 5.2.2). Doing so might put the original author in a difficult position. For example, imagine if a set of interviews had been conducted with members of a local community to better understand their attitudes towards a potential nuclear power plant to be built in that region. Creating a poem that revealed the identities and attitudes of individuals would place them in a difficult situation within both their own community and in relation to other publics (e.g. activists, lobbyists, journalists) as well.

Allow the Poem to Develop

The final core principle helps to determine what should be done with the poem once it has been created. As will be discussed in Sect. 5.2.7, a key principle of this research process is in the sharing of the transcribed poem, the feedback from which might encourage you to re-edit your poetic transcription. It might also be that others choose to edit, shape, and re-mix your poem for their own purposes, and to explore other narratives and lived experiences. Acknowledging that the 'final' poem that you produce prior to sharing it with others is not necessarily the final evolution of the poem, affords both you and others the opportunity to re-visit the poem and its potential impact with various publics.

The method for poetic transcription that I have developed from these core principles can be broken down into the following seven stages, each of which I will discuss in the proceeding sections: formulate research questions, select data, read data, code and categorise, write a poem, confirm trustworthiness, and share.

5.2.1   *Formulate Research Questions*

Determining your research questions is a fairly similar process to that employed in poetic content analysis (see Sect. 4.3.1). Endeavour to develop research questions that are consistent with an interpretivist theoretical perspective, and which cannot be answered with a simple 'yes/no' answer. You should also ensure that these research questions are coherent with the core principles that underpin this research method, as presented in Fig. 5.1. However, unlike with poetic content analysis you are not limited to using poetry as a data set, and this means that there is a much greater degree of freedom with regards to the development of your research questions.

Let us take one of the research questions that was posited in Chap. 4, for use in poetic content analysis:

RQ-PCA   to what extent can poetry be used to better understand how first-year undergraduate students perceive the importance of ethics in science?

Given that poetic transcription is not limited to the sole use of poetry as a data set, this research question could be broadened out considerably:

RQ-PT   how do first-year undergraduate students perceive the importance of ethics in science?

This research question is still consistent with an interpretivist theoretical perspective, it avoids a formulation that leads to a yes/no answer, and it is also well aligned to the four core principles set out in Fig. 5.1.

Exercise 5.1: Formulate a Research Question

Think about a specific science communication topic that you would like to investigate and create one or two research questions for which poetic transcription would be an appropriate method to adopt. As with Exercise 4.1, you should avoid those research questions which could be answered with just a yes/no answer. Make sure that they are compatible with both an interpretivist theoretical perspective and the core principles shown in Fig. 5.1.

5.2.2   Select Data

In terms of selecting the data for your poetic transcriptions there are two broad approaches: either purposefully creating data sets or using secondary sources of data.

In the case of purposefully created data sets, these might take the form of interviews, surveys, field notes, ethnographic observations, or any other textual source. This need not be limited to traditional qualitative data sets, as emails, letters, or even poetry could all be used as data for this method. What is more important is ensuring that the correct ethical consent has been sought and approved, and that anyone who is contributing to the data set is informed of the ways in which it is going to be used, which should also include the concept of the poetic transcription itself (see Chap. 6 for a thorough discussion of how to create such data sets in an ethical manner).

In terms of using secondary sources of data, this again breaks down into two broad types. Firstly, there are data that have been collected for the purpose of another research project, and secondly there are data that exist in the public sphere, but which have not been created for the purpose of a research project. In the case of the former, it is still necessary to check that the data has been collected ethically. If this is not the case or it is ambiguous, then you should avoid using this data set, as to do would be ethically unsound. If the data has been collected and shared in an ethically appropriate manner, then I would still recommend getting in contact with the original researcher(s) who collected the data, and to inform them of your proposed research. Doing so is not only a professional courtesy, but it will also help to provide additional context for the data, highlight any issues that need to be considered (e.g. missing segments, or sections which need to be updated or redacted prior to analysis), and provide you with a useful network for when you come to share the transcribed poem (see Sect. 5.2.7). In terms of actually finding these data, I would recommend looking for qualitative research articles that are related to your research question. Most open access journals also have an open data policy in which data are stored in a repository for other researchers to use. If the research article that you are interested in does not contain a link to such a data set, then you could instead try contacting the lead author to obtain such access.

For data that were not purposefully collected for research, additional steps need to be considered. In the case of using data that were produced for public consumption, but for which the author retains the copyright or intellectual property (e.g. book extracts, newspaper articles, song lyrics) then it is advisable to contact the author (or their representatives) to enquire about the fair use policy of their work. While

many authors might be delighted to discover that their work is being used in this
regard, this should not be assumed, and contacting them in advance ensures that
you are not in breach of any legal requirements when you come to share the results
of your inquiry. There is also the possibility of using data that have been shared
on a public platform, but for which the copyright or intellectual property remains
unknown, uncertain, or ambiguous. Here I am mainly referring to data that have been
posted on social media (e.g. Twitter, blogs, Facebook). In terms of negotiating the
ethical and legal implications of this potentially rich data set, 'Social media research:
a guide to ethics' [9] and 'Using Facebook for Qualitative Research: A Brief Primer'
[10] provide excellent guidance, including worked examples for various research
methods. As a guiding principle, using data that has been posted on Twitter, on a
blog, or on a public (i.e. not secret or closed) Facebook Group is usually ok. However,
in doing so you should always consider the four core principles that are outlined in
Fig. 5.1, and in particular care should be taken to ensure that the author cannot be
identified, and that no harm would come to any of the authors of the data as a result
of this process.

Exercise 5.2: Select Some Data
Take the research questions that you created in Exercise 5.1 and find some data
to help you answer it.

I recommend using a secondary data source from another research project
(e.g. interviews, survey responses, field notes), which you know has been
collected and shared in an ethically appropriate manner. Search for a research
article that is related to your research question and try to obtain any data sets
that were used during its construction. You might also consider contacting the
author(s) of this study to provide additional context and to confirm the ethical
provenance of the data.

5.2.3 Read Data

Having identified the data that you will be using for your poetic transcription, the
next stage is to read it all. Reading the data thoroughly, alongside your research
questions, will help to familiarise yourself with both the text and its author(s). You
needn't take any notes at this stage, although in terms of 'cleaning' the data, if you
come across any identifiable information then this is the point at which it should
be redacted. Redacting such information at this first pass of the data will ensure
that it does not appear in the transcribed poem, and also avoid any such information
contributing to the emergent narratives that you will begin to identify in the next
stage of this process. Cleaning the data might also involve removing any data sets

that are incomplete, ensuring that the data are audited correctly, and potentially filing any gaps (e.g. missing survey responses). Getting to know your data in this manner is an essential part of any qualitative research method; good analysis depends on understanding the data and doing so will help to highlight any limitations in either the data or how it has been collected [11].

Exercise 5.3: Read Some Data
Read the data that you identified in Exercise 5.2, alongside the research questions from Exercise 5.1. Is there any data here that should be redacted because it identifies the author either explicitly or implicitly? If so, then redact it now so that it does not form a part of the subsequent coding process and is not considered for inclusion in the transcribed poem. Clean any other data in terms of missing responses and poor documentation and keep an audit of how and why you have done this.

5.2.4 Code and Categorise

The next stage of this process is to label the data with appropriate codes and to look for emergent categories and narratives. This process is very similar to that which was discussed in Sects. 4.3.4 and 4.3.5 for poetic content analysis, and essentially involves reading through all of the data and assigning codes in relation to your research questions, following which a grouping of categories should be constructed. As with the process discussed in Chap. 4, this should be done until data saturation is reached, any newly emergent codes should be checked against the previously coded data, and a careful audit should be carried out to ensure the trustworthiness of the approach (see Sect. 5.2.6). During the initial coding process (i.e. before categorisation), it is especially important to make a note of any key phrases that best represent each label, as these may later be used to help construct the transcribed poem (see Sect. 5.2.5).

To demonstrate this coding and categorisation, imagine that we are aiming to answer the following research question:

RQ-career   how did scientific researchers chose their particular career pathway?

Suppose that in answering this research question we are using a collection of semi-structured interviews that have been conducted with scientific researchers, published on a blog and for which ethical clearance and permissions have been obtained. One of the questions that all of the interviewees was asked was 'Why did you choose to study science?' A sample response, and its coding, is shown in Fig. 5.2, which is an excerpt from an interview that I gave to the Australian Science Communicators website in 2020 [12].

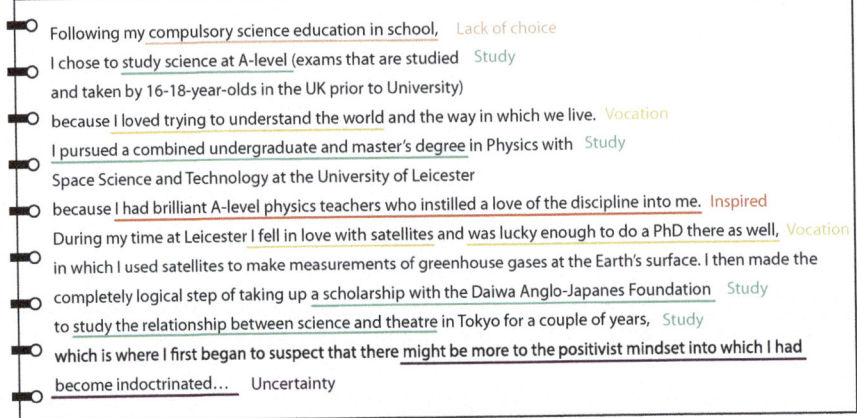

Fig. 5.2 An example of the coding process for poetic transcription, based on an interviewee's response to the question 'Why did you choose to study science?'

The emergent codes are shown in Table 5.1. This table would then continue to be filled following all of the responses from all of the interviewees, with several example sections of text being recorded for ease of reference.

Imagine that interviews with 10 scientific researchers were conducted and then coded in a similar manner. Such an approach might produce a group of categories like those shown in Table 5.2 (this table is only illustrative; it has not been created from any actual data other than that shown in Fig. 5.2). Again, it is important to keep a careful record of any changes to the codes or categories during this process, the representative examples that might be used in the creation of the poem, and potentially the frequency of their occurrence.

Table 5.1 The codes that emerged from the coding of above interview response in relation to RQ-career

Code	Description	Example	Frequency
Inspired	The interviewee was inspired by someone else	'I had brilliant A-level physics teachers.'	1
Lack of choice	The response indicates that the choice was beyond their control	'compulsory science education.'	1
Study	Education played a role in the decision	'pursued a combined undergraduate and master's degree.'	4
Uncertainty	The interviewee displays doubt in their choices	'I had become indoctrinated.'	1
Vocation	There was an emotional response/'a calling'	'I fell in love with satellites.'	2

Table 5.2 The categories that emerged for RQ-career, alongside their corresponding codes

Category	Corresponding codes	Example	Frequency*
Destiny	Vocation, Lack of choice, Accidental	'I fell into it.'	14
Doubt	Uncertainty, Regret	'I had become indoctrinated.'	21
Education	Study, Teachers, School, University, Informal learning	'From the first time my teacher showed us a Bunsen burner'	37
Impact	Society, Solutions, Improvement	'I just wanted to save the world.'	17
Support	Inspired, Advised, Mentored	'They made me believe in what I could achieve.'	41

*the number of occurrences is not limited to one per poem

It is also advisable to write a couple of sentences outlining what is meant by the emergent categories unless this is obvious. The reason for doing so is that unlike for the poetic content analysis shown in Chap. 4, this poetic transcription method will not explicitly analyse each of these categories with respect to the data and other literature, but rather it will use them to help shape the transcribed poem. Keeping a record of what is meant by each of these categories will help you in the formation of the poem, while also improving the trustworthiness of the approach.

For the example shown in Table 5.2, the categories might be represented as follows:

Destiny—Such data demonstrate that the interviewee believed themselves to be destined to pursue a scientific career.

Doubt—These are examples from the data that evidence the doubt that the interviewees have expressed in choosing a scientific career, either now or in the past.

Education—These are segments of the data that refer to education (in either a formal or informal setting) having a significant impact (positive or negative) on the interviewees choosing a scientific career.

Impact—These portions indicate that the interviewee chose a scientific career to make an impact on society, ranging from the very local to the international in terms of scale.

Support—These instances represent examples of support that the interviewee found essential in enabling them to pursue a scientific career.

Exercise 5.4: Code and Categorise the Data

Go through the data that you have selected in Exercise 5.3 and begin to code it according to your research questions. Each time a new code emerges, go back to the start of the data set, and look for additional occurrences. Be sure that data saturation has been reached, to keep a detailed auditing trail, and to make a note of any particularly illustrative or evocative sections of text for each of the codes. Once this has been done, look for any emergent patterns or narratives that could be used to group these codes into broader categories, documenting the process and making note of any particularly demonstrative sections of text.

5.2.5 Write a Poem

You should now have a list of emergent narratives, and several lines that provide illustrative examples for each of these categories. The next step in the poetic transcription process involves turning these lines into a poem that gives voice to these narratives, and which is representative of (and fair to) their original authors. In terms of actually constructing the poem, this will to some extent depend on your own writing process, and it might be guided by a specific style or structure (see Chaps. 2 and 3 for examples). However, I find the method outlined in Fig. 5.3 to be a useful guide for this process and is one that you may wish to adopt.

At each of these stages you should also be guided by the core principles shown in Fig. 5.1. For me personally, I believe that this precludes the addition of any words that do not appear in the text that you are analysing. These core principles also suggest that while trimming the beginning or end of a section of text (providing that this

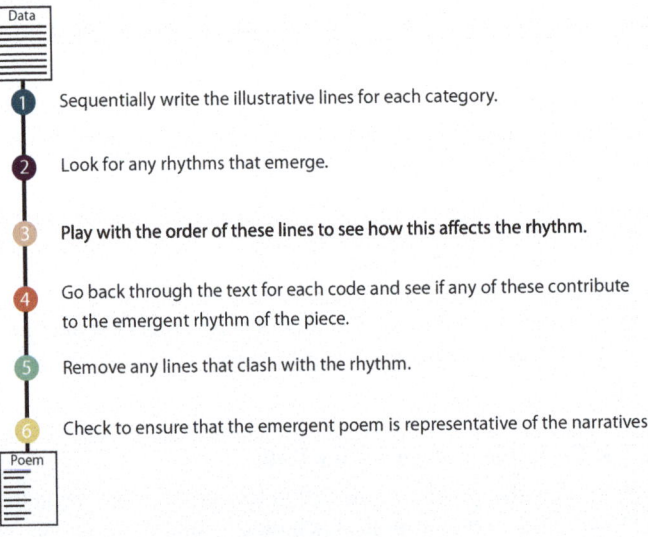

Fig. 5.3 A six-step guide for writing a transcribed poem

does not intentionally change the context) is permissible, cutting out central portions of any text (i.e. merging the beginning and end of a segment, after removing some words in the middle) should be avoided. Adding grammar and correcting for spelling mistakes is also within the spirit of these core principles but can be avoided if to do so would compromise the emergent narratives of the piece.

To demonstrate this approach, let's take the data presented in Table 5.2 and apply the six steps shown in Fig. 5.3.

Step 1: Sequentially write the illustrative lines for each category

Writing out the illustrative lines (in reality there will likely be more than one per category) gives the following:

I fell into it.

I had become indoctrinated.

From the first time my teacher showed us a Bunsen burner.

I just wanted to save the world.

They made me believe in what I could achieve.

Step 2: Look for any rhythms that emerge

There are potentially some nice rhythms at play here, especially the line 'They made me believe in what I could achieve', but at the moment they look a little unwieldy, and the poem itself is lacking any structural narrative.

Step 3: Play with the order of these lines to see how this affects the rhythm

Re-arranging the lines, adding some line breaks, and correcting for grammar then gives the following:

I fell into it

from the first time my teacher

showed us a Bunsen burner,

they made me believe

in what I could achieve.

~~I just wanted~~

~~to save the world.~~

XXX

I had become indoctrinated.

This is already starting to look, sound, and feel much more like a poem—mostly driving by the rhythm that is starting to emerge. However, there are still some parts that are missing and others that need re-visiting. I have crossed out the lines 'I just wanted to save the world' as it feels like a big leap from the previous lines in the poem. However, the 'Impact' category is an important aspect of the emergent narrative and so I need to find some other lines that can replace these. Similarly, the positioning of wanting to 'save the world' next to 'become indoctrinated' potentially recontextualises being indoctrinated into a positive experience, whereas we know from the emergent category of 'Doubt' that this is not the case. As such I have included 'XXX' to indicate that we need some additional lines here to help ensure that this important aspect of the emergent narratives is not lost or mispresented.

Step 4: Go back through the lines for each code and see if any of these contribute to the emergent rhythm of the piece

Going back through the illustrative examples that I have noted for the codes, prior to categorisation, imagine that I find the following four lines:

'But I had found my road'

'Turning questions into answers and answers into hope'

'There's always doubt'

'It's not paved with gold'

The first of these two lines can be combined to replace the lines 'I just wanted to save the world', while the latter two could be combined to replace the 'XXX', to give the following:

I fell into it.
From the first time my teacher
showed us a Bunsen burner,
they made me believe
in what I could achieve.
I had found my road.
Turning questions into answers
and answers into hope.
There's always doubt,
it's not paved with gold –
but I had found my road.

Step 5: Remove any lines that clash with the rhythm

You'll see that I also decided to remove the line 'I had become indoctrinated', and to replace it instead with 'but I had found my road', which provided an effective call-back to the earlier section of the poem and helped to re-position the category of 'Doubt' as being important but not overly dominant in terms of the emergent narrative of the data. I also edited the line 'But I had found my road' when it first appears in this poem, by removing the word 'But', as it improves the rhythm of the piece without affecting the context.

Step 6: Check to ensure that the emergent poem is representative of the narratives

Is this piece consistent with the emergent narratives of the text? While the poem is itself open to interpretation by the reader, each of the five categories ('Destiny',

'Doubt', 'Education', 'Impact', and 'Support') is well represented by the lines of the poem and there is not an overly dominant reading that contradicts the narratives that arose from the coding and categorisation process.

We should also make sure that the transcribed poem, and the process adopted, is well aligned to the four core principles of the poetic transcription method as shown in Fig. 5.1.

- **Does it give voice to the emergent narratives?** Yes. All five emergent categories are evident in the transcribed poem.
- **Has the narrative been given preference to the aesthetic quality, where necessary?** Yes. No additional words have been added, nor has any context purposefully been altered to allow for artistic 'fit'.
- **Does the poem contain any identifiable information?** No. There are no identifying pieces of information either explicit or implicit in nature.
- **Is the poem the final version?** No. I will discuss what might now be done with the poem in Sect. 5.2.7.

Exercise 5.5: Write a Poem
Take the illustrative lines and emergent narratives that you developed in Exercise 5.4 and use them to create a poem. Figure 5.3 can be used as a guide to help you in this process, although you do not have to stick to it rigidly. If you adopt another process, then be sure to keep a detailed record of what you have done and why, to help confirm the trustworthiness of the piece. Remember to also be guided by the core principles shown in Fig. 5.1, and to ensure above all else that you have truthfully represented the voices of the authors whose words you are transcribing.

5.2.6 Confirm Trustworthiness

To confirm the trustworthiness of this method, and the transcribed poem that you have created, much of what was discussed in Sect. 4.3.6 for poetic content analysis still holds true. You should keep a detailed audit of the process, ensure that data saturation has been reached, and fully justify every decision that you made in the coding, categorising, and writing stages. Furthermore, making sure that your approach is aligned to both an interpretivist theoretical perspective and also the four core principles shown in Fig. 5.1 will further strengthen the validity of your transcribed poem. Multiple coders might also be used in this process, although care must be taken at the writing stage to enable equal weight to be given to the transcription and interpretation of the researchers themselves. Finally, the trustworthiness of the transcribed poem is further strengthened by the final stage of this process, and how this is then used to potentially reflect upon and then re-edit the poem that has been created, as will be discussed in Sect. 5.2.7.

Exercise 5.6: Consider Your Trustworthiness
Go back through the coding and categorisation that you have performed in Exercise 5.5. Do you have a detailed audit trail that would enable other researchers to follow what you have done at every stage? How in-depth is your codebook? If you had to start the process from the beginning, would you arrive at the same set of categories given the documentation that you have provided? With regards to the writing of the poem, to what extent have you justified every editorial decision, and are these in line with the core principles shown in Fig. 5.1?

5.2.7 Share

Now that you have the first version of your transcribed poem, and you have confirmed its trustworthiness it is time to share it. Identifying the audience with whom to share the poem will largely be determined by both your research questions and the way in which you have constructed and carried out the poetic transcription process. For example, if the data that you used came from participants who were explicitly recruited for this research question, then you might choose to share it with them. Similarly, if you are making use of a secondary data set, then you could share it with the original auditors of that data set, who in turn might offer to share it with their participants. If you are using data that has been obtained via social media then you could share it with members of that online community, and if instead the data was obtained from the written text of other authors, then depending on the research questions it might be appropriate to share it with either the author themselves or else critics, scholars, or fans of their work.

Once you have identified your target audience, you should be clear in the reasons for your engagement. As well as outlining your research question and rationale (or repeating this if they have already given their informed consent at an earlier stage in the research process), you should also be explicit in the format you wish their feedback to take, and the way in which you plan to use it. Be honest, open, and courteous in your correspondence. The extent to which you take this feedback on board can be difficult to judge, as you need to be representative of the original data, while also being respectful of the participants and/or related communities. Be transparent with the way in which this feedback is used, being sure to document and justify every editing decision in the process. You might opt to do this sharing and editing multiple times and with multiple people; in every instance your approach should be consistent with the four core principles outlined in Fig. 5.1. For example, if someone identifies a line or phrase as being identifiable of either themselves or another author then it should be removed.

Even after several such iterations with the appropriate communities, the transcribed poem that you arrive upon might not be the final version of the poem. For example, upon submission of your poem and process to a peer-reviewed journal, one of the peer-reviewers might identify a participant from the transcribed poem. Alternatively, upon publication a member of another community might decide to adapt the poem based on their own lived experiences. This would not necessarily require you to re-edit the latest version of the poem that you presented, but rather it is testimony to the co-creative nature of this research method and the opportunities for creativity and agency that the four key principles outlined in Fig. 5.1 can afford.

Continuing the illustrative example of the transcribed poem that was presented in Sect. 5.2.5, imagine that this was then presented to the scientists whose interviews were used as the data for this study. Suppose that the majority of them were happy with the poem, and that while several of them felt that it was not specifically representative of their own personal journeys into scientific careers, they understood how the narrative spoke for the wider collective. However, one of the interviewees noted that the lines 'from the first time my teacher showed us a Bunsen burner, they made me believe' were similar to some which they had used on a very popular YouTube video, and as such they were worried that they might be identifiable from the poem. They suggest editing the poem to read as follows:

I fell into it.

From when my teacher

first showed us

cause and effect,

they made me believe

in what I could achieve.

I had found my road.

Turning questions into answers

and answers into hope.

There's always doubt,

it's not paved with gold –

but I had found my road.

This suggestion has not come from the data but removing the original (identifying) line is in keeping with the core principles of this research method. Furthermore, the new line that is suggested is still representative of at least one of the emergent categories ('Support'). After sharing the newly edited poem with the remainder of the interviewees, several of them comment that this is an improvement, as it moves the narrative away from an association with one specific area of science (i.e. chemistry) towards a more inclusive whole. In this instance, sharing the poem, eliciting feedback, and using this in an appropriate manner has thus improved the transcribed poem in terms of how it represents both the emergent narratives of the data and the voices that created that data in the first instance.

Exercise 5.7: Share Your Poem
Identify an audience with whom to share your transcribed poem from Exercise 5.5 and use their feedback to reconsider and potentially re-edit your work. If you have followed the exercises in this Chapter and utilised a secondary data set to answer your research questions, then contact the researcher(s) from that study and ask them to provide feedback. When deciding how to use this feedback to re-edit the poem, refer back to both your original research questions and the four core principles outlined in Fig. 5.1. Be sure to be honest, open, and courteous with regards to how you will use this feedback, and document both the process and any changes that are made to the poem as a result.

The poetic transcription presented in this Chapter (and which is illustrated in Fig. 5.4) is one specific method that I have developed based on the core principles shown in Fig. 5.1 for use in science communication research. I encourage you to both adopt and adapt this method for your own purposes. However, before doing so I would recommend determining your own core principles, so that you can be sure that you have a harmonious and defensible research methodology. For example, you might decide that you want to create a poetic transcript from interviews that were conducted with a named individual, in which case (assuming that these interviews had been obtained in an ethical and informed manner) you could negate the third core principle of my method (i.e. 'The poem should not contain any identifiable information'), as doing so might better enable the voice of the interviewee to be heard.

5.3   In Practice: Moving Scientific Conferences Online

To further illustrate how the poetic transcription shown in Fig. 5.4 can work in practice, I now present a worked example for each of the seven steps, demonstrating how to move from the formulation of research questions, through to the selection and coding of data, and the writing and sharing of a poem.

Fig. 5.4 The seven steps of
the poetic transcription
research method

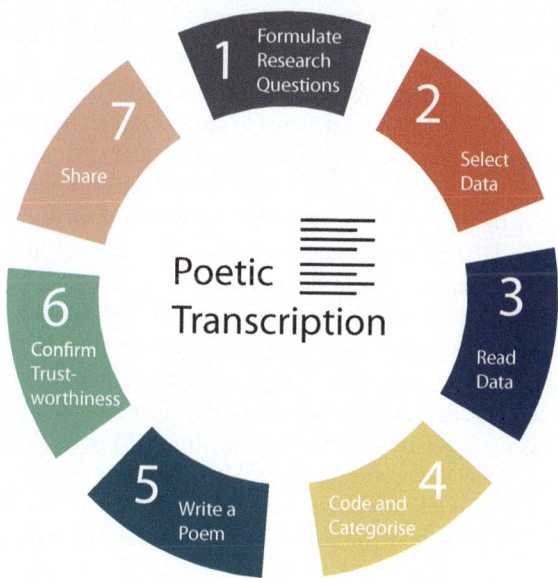

5.3.1   Formulate Research Questions

The impact of COVID-19 upon the scientific community is wide and far-reaching, from research funding [13] and international collaboration [14] to gender inequality [15] and education [16]. One of the most immediate and visible impacts was that which COVID-19 had on academic conferences. One such conference was the European Geosciences Union (EGU) General Assembly, an annual conference for Earth, planetary, and space scientists, that takes place every year in Vienna, Austria, attracting approximately 16,000 participating scientists. The 2020 General Assembly was due to take place from the 3rd to the 7th of May; however, by mid-March it became apparent that the General Assembly could not take place as an in-person event, and so it was announced that it would be cancelled and replaced with an online alternative. This gave the conference organisers approximately six weeks to develop and deliver a virtual event, with 'EGU20: Sharing Geoscience Online' being the first geoscience conference of its size to go fully online. Moving the conference online presented an opportunity to better understand what people enjoyed about the EGU General Assembly, and to potentially capture and analyse any opportunities that were presented as a result of the event going virtual. This was formalised into the following two research questions (RQs):

RQ-EGU1:   what did people miss from a regular General Assembly?
RQ-EGU2:   to what extent did going online impact the event itself, both in terms
           of challenges and opportunities?

5.3.2 Select Data

In terms of selecting the data for this poetic transcription, I will be using a data set that was specially constructed to answer RQ-EGU1 and RQ-EGU2. These data were originally analysed as part of a research study that made use of a traditional approach to qualitative content analysis [17], and I am now making use of these data to perform a poetic transcription. To collect these data a survey was distributed to all attendees of EGU20: Sharing Geoscience Online, which asked them various questions relating to their experiences of the virtual conference. The survey was distributed using the public survey platform of Zoho Forms, and the link to the survey was distributed to all participating scientists via email. The link to the survey was also distributed over social media, using EGU's official Twitter, Facebook, LinkedIn, and Instagram accounts, as well as being shared by various other affiliated accounts, and the survey was open for responses from the 4th of May until the 1st of June 2020. The data was collected according to the British Educational Research Association's (BERA) ethical guidelines for educational research, and before the subsequent stages of this research method, the names and institutes of all participants were removed.

5.3.3 Read Data

Once the survey data had been collated and cleaned of incomplete answers, there were 1,580 responses. The data was then read by both myself and my colleague Hazel Gibson. Given the extremely large dataset it was decided that in the first instance we would each read the same 100 complete responses from a random sample of participants. In addition to familiarising ourselves with the data, any identifying information was also removed so that it would not be included in the subsequent coding and categorisation.

In this particular example, a dual coding approach was adopted until the writing of the poem (Stage 5 in the process; see Fig. 5.4), at which point I created the poem independently of my fellow coder. The reason for this approach will become evident in Sect. 5.3.7, with the use of dual coding helping to further improve the reliability and validity of the emergent codes and categories (see Sect. 5.3.6).

5.3.4 Code and Categorise

Both Hazel and I began by coding the 100 sets of random responses according to RQ-EGU1 and RQ-EGU2. The individual codebooks that we used are shown in Tables 5.3 and 5.4, respectively. Both of us found that by the end of this coding process there were mounting instances of the same codes, but no new ones, and that data saturation had thus been reached. Note that in this instance we did not record the

Table 5.3 The codes that emerged when Hazel conducted a qualitative content analysis of the data including their descriptors and examples of their occurrence

Code	Description	Example
Networking	Indicated missing in-person interactions, contact, and friendship	'Seeing my colleagues and interacting in person'
Multiple Formats Communicating	Referred to different formats for communicating, such as: viewing, discussing, listening, debating, and multiple format communication	'Verbally communicating to people while visually inspecting their work'
Detail	Described details of science, and in-depth conversations	'Without the visual interface it's very difficult to go into details'
Behaviour	Related to the behaviours and attitudes of the participants and those that were expressed by others	'People don't respect their time slots and have cross conversations'
Spontaneity	Indication that the participants missed freedom within the schedule, with time to talk, debate, explain, find unexpected subjects, interactions, or conversations	'Spontaneous questions, time for a more personal, friendly chat'
Preparation	A discussion of the preparation of scientific materials, talks, formats, etc	'Scientifically I could prepare/have more in-depth discussion'
Flexibility	Addresses the possibility of flexible interactions, of being able to move between sessions, and/or of multi-tasking	'Often the whole session is not totally of interest and you would like to change room just for one talk'
Open Access Science	Highlights the issues of open access science, sharing science, and expanding the reach of research	'The impact is undoubtable greater than in classic EGU GA where only a few people could stand in front of poster'
Emotion / Nostalgia	Discusses missing the in-person event, and/or feelings of an intangible sadness, excitement, joy, or boredom	'Everything! Nothing can replace the face-to-face event'
Overcoming Current Events	Raises the issue of overcoming non-specific challenges of COVID-19 to carry on with plans regardless	'You did an amazing job in a short time, and considering the current situation in the world'
Attendance	The participants state if they are able to attend or not attend the virtual conference, despite their original plans for the in-person event	'It has allowed me to attend a meeting I could not attend in the first place'

(continued)

Table 5.3 (continued)

Code	Description	Example
Waste of time	The responses indicate that the virtual event was a waste of time and/or a disappointment, and that it would have been better off cancelling it	'I don't see the point of this format, EGU had better been completely cancelled'

Table 5.4 The codes that emerged when I conducted a qualitative content analysis of the data including their descriptors and examples of their occurrence

Code	Description	Example
Deeper engagement	These responses indicate that these participants were able to have a deeper engagement in terms of either more questions or longer discussions etc	'Scientifically I could prepare/have more in-depth discussion'
Good for Early Career Scientists	The responses highlight that the virtual event presented good opportunities for Early Career Scientists	'During oral presentations, generally time for questions is very narrow, and you do not always feel it is your place to do so as an ECR. Having this ability during the whole session time slot is really enjoyable'
Difficulties with Tech	Participants encountered difficulties accessing the online content	'The chat pages have some glitches. Comments sometimes disappearing for unknown reasons in my window, while other people could see them'
Networking	Participants missed the opportunity to professionally network in person	'Meeting people! Networking! The chat it great but it is just not the same'
Socialising	Participants missed the opportunity to catch up with colleagues and friends in person	'I can't see my teachers and classmates, we can't talk questions face to face, sometimes, the text-chat can't arrive the effect. And I miss the scenery and food of Austria, haha'
Too much info	Participants felt overwhelmed with the number of communications they received	'The emails where too long and un-structured, plus a bit spammy (emails as author, co-author, personal program, convener....)'
Lack of engagement	These responses indicate that the virtual format presented fewer opportunities for deep engagement on scientific topics	'The 15-min orals and as long as need discussion for the posters. This format cuts down on the ability to explain, drastically. I don't think it's been translated good enough'

(continued)

Table 5.4 (continued)

Code	Description	Example
Environment	Participants found that the virtual conference had a positive impact on the environment, in comparison to an in-person event	'Carbon footprint issue. Obviously, we do not need to go every year to such meetings. So remotely following them is very interesting. And if you have personal restrictions (accessibility, money, childcare) preventing you to attend, that's quite an improvement!'
Boring	The virtual event was found to be less vibrant than the in-person meeting	'Nothing special and there are plenty of ways to explore to make this feel more interactive. Scrolling through the presentations makes attendance feel a lot like grading papers'
Convenience	Participants found the virtual event more convenient to attend	'Reduce long distance transportation while maintaining the visual and verbal aspects'
Lack of info	These responses indicated that participants found it difficult to 'discover' the conference or find out how to attend specific webinars etc	'Found it hard to access the talks or find info about how to attend webinars but the rest was well advertised'
Inaccessible	Participants found the virtual format to be inaccessible	'I can't concentrate on the virtual meeting, although it's great, especially in text-chat section, I can't follow other people's idea'
Accessible	Participants found the virtual format to be more accessible than an in-person event	'Those unable to physically attend can gain some part of the experience from home. That includes physically disabled and financially unable'
Discovery	Participants noted that the virtual nature of the conference made it more likely to have 'accidental discoveries' in comparison to the in-person event	'Also people who normally might not ask questions in person would be more inclined to do so online'

frequency of occurrence for the codes, as we were more interested in ensuring that we had captured all emergent narratives rather than the regularity of their appearance.

After this initial coding exercise was completed, Hazel and I combined our code-books and decided on a number of categories that covered all of these codes, which could in turn be used to better represent the emergent narratives. These combined categories are shown in Table 5.5 alongside the original codes and coder (HG is Hazel Gibson; SI is Sam Illingworth).

Table 5.5 The initial combined categories and descriptions that were used to group the codes shown in Tables 5.3 and 5.4

Category	Description	Codes (original coder in brackets)
Information	How participants were informed of the new format, and how they accessed this information	Attendance (HG), Waste of Time (HG), Difficulties with Tech (SI), Too much Info (SI), Lack of Info (SI)
Connecting	How networking and socialising were impacted by moving to a virtual conference	Networking (HG), Networking (SI), Socialising (SI)
Engagement	The extent to which the virtual environment either encouraged or restricted engagement. Also includes spontaneity/discovery of sessions	Multiple Format Communicating (HG), Spontaneity (HG), Preparation (HG), Emotion / Nostalgia (HG), Deeper Engagement (SI), Lack of Engagement (SI), Boring (SI), Discovery (SI)
Environmental Impact	How changes to a virtual conference impacted the environment	Overcoming Current Events (HG), Environment (SI)
Accessibility	The extent to which a virtual conference was more or less accessible to different audiences	Detail (HG), Behaviour (HG), Flexibility (HG), Open Access Science (HG), Convenience (SI), Inaccessible (SI), Accessible (SI)
Early Career Scientists	The impact that the virtual environment had on Early Career Scientists	Good for Early Career Scientists (SI)

Following this initial coding and categorisation, Hazel and I realised that some of the categories shown in Table 5.5 did not fully relate to RQ-EGU1 and RQ-EGU2, and so it was decided that the 'Information' and 'Early Career Scientists' categories should be removed. The former because these responses were more concerned with technical changes and difficulties, and the latter because it was decided that it would be discriminatory to highlight one specific demographic of researchers. Table 5.6 presents the final categories to emerge from these data in relation to the research questions, alongside several examples.

To confirm the trustworthiness of this approach 50 random respondents from each of three distinct demographic groups (Early Career, Mid-Career, and Senior Career scientists; all of which were self-identified by the participants as part of the survey) were selected. Hazel and I then individually assigned the categories shown in Table 5.6 to the responses for these 150 participants and independently observed that there was no newly emergent codes or categories. This additional step helped to provide confidence that the categories shown in Table 5.6 were the correct emergent narratives with which to construct the poem.

Table 5.6 The final categories to emerge from the data in response to RQ-EGU1 and RQ-EGU2, alongside a number of examples for each category

Category	Definition	Examples
Connecting	How networking and socialising were impacted by moving to a virtual conference	'The chat it great but it is just not the same' 'My job as a scientist is mostly reading and writing, the physical conference is breaking out of this, which opens many other opportunities to think, cooperate, and pathways to discuss'
Engagement	The extent to which the online environment either encouraged or restricted engagement. Also includes the spontaneity/discovery of sessions	'Maybe I come from an old school, but attending a conference directly offers many possibilities to establish contacts with other scientists, to interact in a deeper and less aseptic way than online event provides' 'Nothing can replace the face-to-face event' 'Scrolling through the presentations makes attendance feel a lot like grading papers' 'It may be topic related, but this time was the first time that I got exactly the kind of feedback to my presentation I was hoping for' 'I could take part in sessions at the fringe of my expertise since the short summaries given by presenters helped me to understand their core message' 'I'm concerned about the copyright issues when uploading presentation'
Environmental Impact	How changes to a virtual conference impacted the environment	'As geologists we really need to think about being more climate-friendly in our jobs!' 'Because the environmental footprint of normal EGU seems unreasonable nowadays, we have to think differently and this crisis pushes a bit too far but shows us alternatives' 'If it was only online, we'd have to adapt to a new way of working, which would ultimately accelerate our transition to a green future' 'The traditional conference is getting more difficult to justify with climate change and the requirement that everyone jet around the world to discuss earth science, especially science related to climate change'

(continued)

Table 5.6 (continued)

Category	Definition	Examples
Accessibility	The extent to which a virtual conference was more or less accessible to different audiences	'Those unable to physically attend can gain some part of the experience from home' 'I think the online format allowed people who could not come to the meeting for cost or travel restrictions to attend, thus broadening the scientific content' 'The expanded attendance is good, but there is definitely something lost: but also something gained (accessibility)'

5.3.5 Write a Poem

In transcribing the poem I will now diverge from the analysis that was carried out in the original qualitative content analysis of this study. Instead of contextualising the categories shown in Table 5.6 with other relevant research literature (which is the process that was adopted in the original study), I will instead use these to help construct the poem. In doing so, I will adopt the six-step approach shown in Fig. 5.3.

Step 1: Sequentially write the illustrative lines for each category

Writing each of the example lines from Table 5.6 sequentially results in the following:

The chat it great but it is just not the same.

My job as a scientist is mostly reading and writing, the physical conference is breaking out of this, which opens many other opportunities to think, cooperate, and pathways to discuss.

Maybe I come from an old school, but attending a conference directly offers many possibilities to establish contacts with other scientists, to interact in a deeper and less aseptic way than online event provides.

Nothing can replace the face-to-face event.

Scrolling through the presentations makes attendance feel a lot like grading papers.

It may be topic related, but this time was the first time that I got exactly the kind of feedback to my presentation I was hoping for.

I could take part in sessions at the fringe of my expertise since the short summaries given by presenters helped me to understand their core message.

I'm concerned about the copyright issues when uploading presentation.

As geologists we really need to think about being more climate-friendly in our jobs!

Because the environmental footprint of normal EGU seems unreasonable nowadays, we have to think differently and this crisis pushes a bit too far but shows us alternatives.

If it was only online, we'd have to adapt to a new way of working, which would ultimately accelerate our transition to a green future.

The traditional conference is getting more difficult to justify with climate change and the requirement that everyone jet around the world to discuss earth science, especially science related to climate change.

Those unable to physically attend can gain some part of the experience from home.

I think the online format allowed people who could not come to the meeting for cost or travel restrictions to attend, thus broadening the scientific content.

The expanded attendance is good, but there is definitely something lost: but also something gained (accessibility).

Step 2: Look for any rhythms that emerge

Looking at this passage, there are some strong lines and phrases that really stand out. In particular 'The chat it great but it is just not the same' and 'something lost: but also something gained' have a great resonance and will work well together, either side-by-side or to envelop the poem or a section within. Similarly, the lines 'The traditional conference is getting more difficult to justify' and 'Scrolling through the presentations makes attendance feel a lot like grading papers' are really powerful lines that strongly encapsulate two or three of the main categories to emerge from the data ('Environmental Impact' plus 'Connecting' and 'Engagement', respectively).

On the other hand lines such as 'I think the online format allowed people who could not come to the meeting for cost or travel restrictions to attend, thus broadening the scientific content' and 'I could take part in sessions at the fringe of my expertise since the short summaries given by presenters helped me to understand their core message' feel a little unwieldy and the categories which they characterise (i.e. 'Accessibility' and 'Engagement') are better represented elsewhere. Finally, the line 'Nothing can replace the face-to-face event' potentially lends itself to repetition, as it helps to frame the narratives of the emergent categories as evidenced in Table 5.6, i.e. that the virtual conference was unlike any previous in-person EGU General Assembly, but that this was not necessarily a bad thing.

Step 3: Play with the order of these lines to see how this affects the rhythm

Playing with the order of the poem, to foreground the most impactful lines for the four emergent categories, while also correcting for spelling and grammar, and introducing some line breaks resulted in the following:

Nothing can replace the face-to-face event,

the chat it great

but it is just not the same.

My job as a scientist

is mostly reading and writing,

scrolling through the presentations

makes attendance feel a lot

like grading papers.

Nothing can replace the face-to-face event,

but this time was the first time

that I got exactly the kind of feedback

to my presentation I was hoping for.

XXX—AAA

I could take part in sessions

at the fringe of my expertise.

Nothing can replace the face-to-face event,

I'm concerned

about the copyright issues

about being more climate-friendly in our jobs

XXX—Negative

XXX—Positive

Nothing can replace the face-to-face event,

those unable

to physically attend

can gain some part

of the experience from home,

thus broadening

the scientific content.

Nothing can replace the face-to-face event,

if it was only online,

we'd have to adapt

to a new way of working,

which would ultimately accelerate

our transition to a green future

The chat it great

but it is just not the same,

there is definitely something lost

but also something gained.

The line 'XXX—AAA' acknowledges that an additional line is needed here which highlights some of the positives of the online event in terms of 'Engagement'. Similarly, the lines 'XXX—Negative' and 'XXX—Positive' refer to the need to provide additional examples for negative and positive experiences of the online event, ideally starting with the word 'about' to aid the rhythm of the piece.

Step 4: Go back through the lines for each code and see if any of these contribute to the emergent rhythm of the piece

To fill in the missing lines of the poem I went back through all of the individually coded survey responses. In searching for a suitable line for 'XXX—AAA' I identified the following quote: 'This made it much easier to think about the contents without the stress of everything around you in the conference centre', the second half of which was chosen as an example from the data that was both illustrative of the emergent category of 'Engagement' and which also fit the rhythm of the poem.

Utilising a similar approach resulted in substituting 'XXX—Negative' for 'about what was possible for presenters' and 'XXX—Positive' for 'about issues that are in common interest'. Also, during this process, the line 'about our conference practices' was found in the data, and I thought it was suitable as a final line for the third stanza, helping to encapsulate the overall emergent narrative of both the data and the poem, i.e. that moving to an online environment enabled many of the participants to reflect on the role and implementation of scientific conferences.

These changes resulted in a version of the poem which now read as follows:

Nothing can replace the face-to-face event,

the chat it great

but it is just not the same.

My job as a scientist

is mostly reading and writing,

scrolling through the presentations

makes attendance feel a lot

like grading papers.

Nothing can replace the face-to-face event,

but this time was the first time

that I got exactly the kind of feedback

to my presentation I was hoping for.

without the stress of everything

around you in the conference centre

I could take part in sessions

at the fringe of my expertise.

Nothing can replace the face-to-face event,

I'm concerned

about the copyright issues

about being more climate-friendly in our jobs

about what was possible for presenters

about issues that are in common interest

about our conference practices.

Nothing can replace the face-to-face event,

those unable

to physically attend

can gain some part

of the experience from home.

Thus broadening

the scientific content.

Nothing can replace the face-to-face event,

if it was only online

we'd have to adapt

to a new way of working,

which would ultimately accelerate

our transition to a green future.

The chat it great

but it is just not the same,

there is definitely something lost

but also something gained.

Step 5: **Remove any lines that clash with the rhythm**

Reading the poem (both out loud and in my head), it is apparent that the last stanza is perhaps a little long. Removing the following lines should help to improve the rhythm of the poem without detracting from the emergent narratives: 'which would ultimately accelerate // our transition to a green future'. Additionally, the line 'about being more climate-friendly in our jobs' feels a bit clunky and could be shortened to 'about being more climate-friendly'.

On reflection, the entire second stanza can also be removed, as it does not particularly add to the poem in any rhythmic sense. It also risks framing the overall piece as being mostly positive, which is not entirely true (see Step 6 below). Finally, switching the third and fourth stanza also seemed to improve the rhythm, without affecting the emergent narratives.

With all of these edits, the poem now has the following structure:

Nothing can replace the face-to-face event,

the chat it great

but it is just not the same.

My job as a scientist

is mostly reading and writing,

scrolling through the presentations

makes attendance feel a lot

like grading papers.

Nothing can replace the face-to-face event,

those unable

to physically attend

can gain some part

of the experience from home.

Thus broadening

the scientific content.

Nothing can replace the face-to-face event,

I'm concerned

about the copyright issues

about being more climate-friendly

about what was possible for presenters

about issues that are in common interest

about our conference practices.

Nothing can replace the face-to-face event,

if it was only online

we'd have to adapt

to a new way of working.

The chat it great

but it is just not the same,

there is definitely something lost

but also something gained.

Step 6: Check to ensure that the emergent poem is representative of the narratives

In this instance I was also involved in the original study from which these data were obtained [17], meaning that I could cross reference the original analysis with the transcribed poem to check for representative narratives with some degree of confidence. For instance, the key message to emerge from the original analysis of the data was that the role of the scientific conference needs to be interrogated with respect to what was learnt from being forced into a virtual environment, i.e. it is not enough to simply return to a 'pre-COVID, business-as-usual' approach. From the transcribed poem that has been presented above, I believe that this narrative is apparent, and that no further editing is required. However, as will be discussed in Sect. 5.3.7, it is only by sharing the poem that I can be confident that this is indeed the case.

I should also make sure that the transcribed poem, and the process adopted, is well aligned to the four core principles shown in Fig. 5.1.

- **Does it give voice to the emergent narratives?** Yes. It is compatible with all of the emergent categories that are shown in Table 5.6, and also with the interpretation of these categories as presented in the original study from which the data was derived.
- **Has the narrative been given preference to the aesthetic quality, where necessary?** Yes. While the rhythm of the piece has been rightfully considered, no edits were made that misrepresented the narratives.
- **Does the poem contain any identifiable information?** No. As far as I can be aware, there is no explicit or implicit identifying information in the poem.
- **Is the poem the final version?** No. See Sect. 5.3.7.

5.3.6 Confirm Trustworthiness

At each stage of the qualitative content analysis that was adopted in this study, the individual codes and categories were re-examined to confirm that they accurately captured the responses of the survey in relation to the research questions. Both Hazel and I carried out this coding independently, until descriptive saturation had been reached [18]. Similarly, a combination of systematic sampling, constant comparison, and detailed auditing and documentation were used to ensure both the reliability (i.e. the consistency with which this analysis would produce the same results if repeated) and the validity (i.e. the accuracy or correctness of the findings) of this approach [19]. Furthermore, after the coding and categorisation had taken place, we shared our preliminary results and methodology with a third independent researcher, Susanne Buiter, who confirmed both the rigour and suitability of our approach.

As is evident from the first draft of the transcribed poem shown in Sect. 5.3.5, its creation was underpinned by the first three core principles shown in Fig. 5.1 (i.e. it gave voice to the emergent narratives, was guided by the data not the aesthetics, and contains no identifying information). Section 5.3.7 also demonstrates the extent to which this process was also compatible with the fourth core principle (i.e. the transcribed poem at the end of Sect. 5.3.5 is not the final poem developed for this study). The alignment with these four core principles helps to build further trust in both the transcribed poem as a response to the research questions, and the process by which it was created.

5.3.7 Share

Having created the transcribed poem shown in Step 5 of Sect. 5.3.5, the final stage in the process was to share it with an appropriate audience. In this instance, the audience that I chose to share the poem with were the co-authors of the original study from which the data was obtained [17], i.e. Hazel Gibson and Susanne Buiter. I chose to share the poem with Hazel and Susanne, as I wanted to be sure that the emergent narratives were consistent with the findings of the traditional content analysis that was performed in this study.

I contacted both Hazel and Susanne and explained the context of the poetic transcription that I was performing, as well as a brief overview of the method that I used. I also asked for feedback on the poem, specifically requesting them to notify me of any elements that misrepresented either the data or the findings of the original analysis. I now present the verbatim feedback that I received, with the kind permission of both Hazel and Susanne to share their words in this format:

Susanne: 'I like the poem. The feedback I can give is really only minor. At first, I thought the 'it' in the second line was a typo ('the chat it great') but then I saw it was repeated at the end, so I guess this is what the participant wrote. The first, third, and last sections bring points of concern (attention) for an online meeting, the second gives a positive (added) aspect, but overall for me the poem reads more towards concern than possibilities. Which of course fits with the'Nothing can replace the face-to-face event'. I felt that the text of the second section was somewhat in contrast with the first line. Nothing can replace the face-to-face event, but it then describes something added. Anyway, minor as I wrote.'

Hazel: 'The thing that really struck me was the way that poem itself gives you the sense of emotional disengagement that was one of the key findings in the study. By using these particular lines in the way you have, the poem felt removed from me as the reader, as though it itself was separated behind a digital barrier. The only thing that I didn't get a sense of was the impact of a remote conference on the different career stages.'

Dealing with Susanne's comments first, I went back and checked the data and found that this was indeed very likely a typographical error by the participant. The original response in full read 'Meeting people! Networking! The chat it great but it is just not the same.', this would suggest that the line in the poem should actually read 'the chat is great' and not 'the chat it great' and changing it to read as such is still in keeping with the four core principles as shown in Fig. 5.1. The remainder of Susanne's comments would suggest that the central narrative to emerge from the poem is indeed consistent with the findings of the original study and the data more generally, i.e. that the virtual meeting raised questions about the way in which scientific conferences could continue to be operated in the future, post-pandemic.

With regards to Hazel's comments, one of the key findings from the original study was that the career stage of the scientists likely had an impact on the ways in which they were affected by the move to a remote conference. For example, early career researchers (ECRs) felt as though they had less opportunities for the networking (both formal and informal) that is essential for establishing both future collaborations and potential job opportunities. Similarly, some ECRs also found the digital format to be liberating in terms of engaging with the speakers and other attendees. Tables 5.3 and 5.4 evidence some of the nuances of the responses with respect to career stage, but these are clearly not evident from the poem. One of the ways with which to address this would be to refer to the variability of experience using some of the participant responses, e.g. 'you do not always feel it is your place to do so as an ECR' or 'maybe I come from the old school', the former of which in particular helps to highlight the individuality of how each participant experienced the transition from the more familiar face-to-face experience.

Re-visiting the transcribed poem in light of Hazel and Susanne's comments results in the following version of the poem:

Nothing can replace the face-to-face event,

the chat is great

but it is just not the same.

My job as a scientist

is mostly reading and writing,

scrolling through the presentations

makes attendance feel a lot

like grading papers.

Nothing can replace the face-to-face event,

those unable

to physically attend

can gain some part

of the experience from home.

Thus broadening

the scientific content.

Nothing can replace the face-to-face event,

I'm concerned

about the copyright issues

about being more climate-friendly

about what was possible for presenters

about issues that are in common interest

about our conference practices.

Nothing can replace the face-to-face event,

if it was only online

we'd have to adapt

to a new way of working.

You do not always feel

it is your place to do so;

the chat is great

but it is just not the same.

there is definitely something lost

but also something gained.

For the purposes of this Chapter, this is where I will end the poetic transcription. It presents a response to the original research questions ('RQ-EGU1: what did people miss from a regular General Assembly?'; 'RQ-EGU2: to what extent did going online impact the event itself, both in terms of challenges and opportunities?') using data that was collected in an ethical manner, the transcription of which has been constructed using the four core principles shown in Fig. 5.1, and ratified by the co-authors of the original study. As demonstrated in Hazel and Susanne's responses, it is also clear that this poem offers something different to the original analysis that was performed on this data. The poem creates an alternative lens through which to listen to the responses of the participants and in turn to experience the challenges and opportunities of moving to a virtual event.

This is, however, not the end of this poem's journey. Other audiences that I could share this poem with include: the original participants of the study (provided they had shared their contact details and given consent to be contacted in this manner), other EGU members, and the wider scientific community. The audience with which the poem is shared would determine both the mode of sharing and the framing of the subsequent feedback. For example, EGU members could potentially be engaged by sharing the poem via EGU's newsletter or social media channels, with members asked to provide feedback with regards to how representative the poem was of their own experiences of EGU20: Sharing Geoscience Online.

The poem could even be presented alongside the original study as an alternative method of generating dialogue about its findings. Alternatively other EGU members or geoscientists might chose to use the poem for their own creative endeavours or to encourage alternative reflections on the role of the scientific conference in a post-COVID world. Acknowledging that the transcribed poem is not the end of the process is an essential part of this research method, as it enables the impact of the findings to reach beyond any one particular researcher or study.

5.4 Summary

This Chapter has introduced you to poetic transcription as a qualitative research method. Beginning with the core principles on which it is formed, I have outlined a specific research method for using poetic transcription in science communication research, highlighting the importance of ensuring that the transcribed poem is truly representative of the data and its author(s). I have also discussed how and why to share the transcribed poem, and the need to treat this as a process which has the potential for constant evolution. By presenting a worked example I have also demonstrated how this seven-step research method can be used in practice. By the end of this Chapter you should have a good understanding of how to use poetic transcription in your own science communication research, and if you have worked through the exercises then you will already be well on your way to developing the skills and processes that are needed to make the most of this versatile research method.

5.5 Suggested Reading

Unlost [20] is an online poetry journal that celebrates the poetry in the existing and the everyday, publishing a wide variety of found poetry that helps to showcase the versatility of this form of poetry. For a comprehensive introduction to poetic inquiry as a qualitative research method you should look no further than *Poetic Inquiry: Craft, Method and Practice* by Sandra L. Faulkner [1], which includes several varied examples of poetic inquiry as well as exercises for developing and using poetry as a qualitative research method. Similarly, 'Poetic transcription with a twist' [21] presents an alternative method to that which been discussed in this chapter; it involves co-creating poetry as a group, using written reflections from the participants as the initial data set for the poetic transcription. This method shifts the power dynamic, as instead of the researcher representing what was shared with them by the research participants (as is the case with the method presented in this Chapter), the research participants themselves construct the poetry from their own words and testimonies. Such a research method is an effective way to ensure that participants are truly represented by the process and would work well as an exercise in the poetry workshops that I will now go on to discuss in Chap. 6.

5.6 Further Study

The further study section in this Chapter is designed to help you practice using poetic transcription as a research method, and to consider how you might use it to develop and interrogate your own science communication research questions.

1. **Define your own core principles.** What are the underlying principles that you think should define your own poetic transcriptions? Are they different to the ones that I presented in Fig. 5.1? Is there anything in my core principles that you disagree with, or which you think might be missing? Reflect on your own theoretical perspectives and research ideologies and use these to create your own set of core principles for poetic transcription.
2. **Adapt the research method.** Having determined what your core principles are, use this to modify the seven-step research method shown in Fig. 5.4. It might be that you keep the same seven steps but apply them in a slightly different manner (e.g. through the use of identifying information, where ethically appropriate), or alternatively you might decide to negate some of these steps or even add your own. Whatever changes you decide to implement, make sure that they are fully compatible with the core principles that you have used to derive them.
3. **Formulate a new research question.** Work your way through each of the steps of your newly adapted research method. How does the final result compare to the transcribed poem that you created using the other exercises in this Chapter?

References

1. Faulkner SL (2019) Poetic inquiry: craft, method and practice. Routledge, New York
2. Prendergast M (2006) Found poetry as literature review: research poems on audience and performance. Qual Inq 12(2):369–388. https://doi.org/10.1177/1077800405284601
3. Pithouse-Morgan K (2016) Finding my self in a new place: exploring professional learning through found poetry. Teach Learn Prof Dev 1(1):1–18
4. Glesne C (1997) That rare feeling: re-presenting research through poetic transcription. Qual Inq 3(2):202–221. https://doi.org/10.1177/107780049700300204
5. Richardson L (1992) The consequences of poetic representation. In: Ellis C, Flaherty MG (eds) Investigating subjectivity: research on lived experience. SAGE Publishing, Newbury Park
6. Prendergast M (2009) "Poem is what?" Poetic inquiry in qualitative social science research. Int Rev Qual Res 1(4):541–568. https://doi.org/10.1525/irqr.2009.1.4.541
7. Faulkner SL (2018) Crank up the feminism: poetic inquiry as feminist methodology. Humanities 7(3):85. https://doi.org/10.3390/h7030085
8. Illingworth S, Bell A, Capstick S et al (2018) Representing the majority and not the minority: the importance of the individual in communicating climate change. Geosci Commun 1(1):9–24. https://doi.org/10.5194/gc-1-9-2018
9. Townsend L, Wallace C (2016) Social media research: A guide to ethics. University of Aberdeen, Aberdeen
10. Franz D, Marsh HE, Chen JI et al (2019) Using Facebook for qualitative research: a brief primer. J Med Internet Res 21(8):e13544. https://doi.org/10.2196/13544
11. Renner M, Taylor-Powell E (2003) Analyzing qualitative data. University of Wisconsin-Extension Cooperative Extension, Madison
12. Australian Science Communicators (2020) ASC Scope Interview: Dr Sam Illingworth. https://www.asc.asn.au/blog/2020/11/05/asc-scope-interview-dr-sam-illingworth-senior-lecturer-in-science-communication-school-of-biological-sciences-uwa. Accessed 10 December 2021
13. Prudêncio M, Costa JC (2020) Research funding after COVID-19. Nat Microbiol 5:986. https://doi.org/10.1038/s41564-020-0768-z
14. Lee JJ, Haupt JP (2021) Scientific globalism during a global crisis: research collaboration and open access publications on COVID-19. High Educ 81:949–966. https://doi.org/10.1007/s10734-020-00589-0
15. Staniscuaski F, Reichert F, Werneck FP et al (2020) Impact of COVID-19 on academic mothers. Science 368(6492):724. https://doi.org/10.1126/science.abc2740
16. Usak M, Masalimova AR, Cherdymova EI, Shaidullina AR (2020) New playmaker in science education: covid-19. J Balt Sci Educ 19(2):180185. https://doi.org/10.33225/jbse/20.19.180
17. Gibson H, Illingworth S, Buiter S (2021) The future of conferences: lessons from Europe's largest online geoscience conference. Geosci Commun 4(3):437–451. https://doi.org/10.5194/gc-4-437-2021
18. Lambert VA, Lambert CE (2012) Qualitative descriptive research: an acceptable design. Pac Rim Int J Nurs Res 16(4):255–256
19. Leung L (2015) Validity, reliability, and generalizability in qualitative research. J Fam Med Prim Care 4(3):324–327. https://doi.org/10.4103/2249-4863.161306
20. Unlost (2021) Unlost journal. https://unlostjournal.com/. Accessed 10 December 2021
21. Smart F (2017) Poetic transcription with a twist: an approach to reflective practice through connection, collaboration and community. Innov Educ Teach Int 54(2):152–161. https://doi.org/10.1080/14703297.2016.1258323

Chapter 6
Poetry Workshops

6.1 Introduction

So far in this book I have largely focused on using poetry as an effective form of science communication via dissemination. Chapters 2 and 3 discussed how you might read, analyse, and write your own science poetry, while Chaps. 4 and 5 presented two research methods for using poetic inquiry to interrogate both science and science communication. In this chapter and Chap. 7 I will move onto how poetry can be used as a media through which to develop dialogue in science communication, i.e. conversations in which knowledge flows between scientific and non-scientific audiences.

In this Chapter, I will discuss how to develop, deliver, and evolve co-creative poetry workshops, followed by a discussion of collaborative projects between scientists and poets in Chap. 7. This chapter will thus centre on using poetry as a facilitatory tool to engender dialogue between scientists and non-scientists, while Chap. 7 will discuss how (and why) to create effective partnerships between scientists and poets.

Engaging non-scientific audiences in dialogues is fundamentally important to the advancement of both scientific research and discourse. Such dialogues, through which non-scientific audiences are given a platform to interrogate the impact and rationale of scientific findings, help to avoid myopia in scientific research, ensuring that it is not conducted in a series of diminishing echo chambers [1]. Diversifying scientific discourse in this way also helps to challenge stereotypes of 'what scientists look like' and 'who science is for' [2]. These dialogues can take many forms [3], from focus groups [4] and citizen science initiatives [5] to citizen assemblies [6] and consensus conferences [7]. The benefits of such dialogues are multidirectional, and as such should be seen as more than simply a 'box-ticking' exercise for engagement.

Imagine, for example, that a group of scientists who are working on flood risk mitigation strategies plan on engaging a local community to discuss plans for a new flood defence system. This community will likely contain members who have been living in the region for several decades, and as such will have expertise that is essential for the scientists to fully understand what defences will work best and

© The Author(s), under exclusive license to Springer Nature Switzerland AG 2022
S. Illingworth, *Science Communication Through Poetry*,
https://doi.org/10.1007/978-3-030-96829-8_6

why. This local knowledge might include awareness of specific regions that are prone to flooding, historic measurements of rainfall that predate official records, and first-hand accounts of major flooding events. This community will also be able to provide expertise with regards to which local residents are likely to oppose any proposed flood defence measures, why this might be the case, and what steps could be taken to encourage them to reconsider their position. By engaging this community in meaningful dialogue, the scientists can use local knowledge to modify their proposals so that they are likely to be more effective. Furthermore, doing so helps to grant agency to the local residents, meaning that they are far more likely to actively support both current and future flood defence measures.

Unfortunately, whenever such dialogue is initiated between scientists and non-scientists there is the danger that hierarchies of intellect might act to stifle any genuine exchange of knowledge [8]. These hierarchies can be established when non-scientific audiences are made to feel like 'non-experts', or that their knowledge and expertise is somehow lesser than that of the scientific 'experts'. This is especially likely to happen when non-scientific audience are drawing on tacit knowledge and/or lived experiences.

Collaborative poetry workshops, in which scientists and non-scientists read, analyse, and write poetry together, help to level these hierarchies. They do so in three ways: by giving permission to the non-scientists to share their expertise, by giving permission to the scientists to display an element of pathos, and by creating a shared sense of vulnerability between participants [9].

By creating poems about a specific subject matter and then sharing the poem, the participants in these poetry workshops are able to express their knowledge and opinions without fear of appraisal; it is harder to be objectively critical of a poem than it is of a belief or judgment (no matter how well informed these may be). Furthermore, reading, analysing, and writing poetry enables participants to explore emotional aspects of a subject while practising the technique of self-distancing when describing difficult events [10]. Participants are then often more likely to engage in dialogue around these subjects.

Encouraging scientists to share an element of pathos (i.e. an appeal to the emotions) helps to challenge stereotypes that scientists are somehow 'cold-hearted' or separated from society [11]. Similarly, sharing poetry in collaborative settings grants scientist the opportunity to explore their own personal perspectives in a way that is not usually afforded to them by more traditional means of scientific dissemination (e.g. scientific articles, conference proceedings, poster displays).

Finally, by creating a sense of shared vulnerability, collaborative poetry-writing workshops help to create a 'safe space' in which respectful dialogue can be nurtured [12]. Once the audience has heard each other share poems that are either deeply personal or aesthetically awkward (or both), a sense of community is developed. As will be discussed in Sect. 6.3, the aesthetics of any poetry that is created are largely unimportant. Rather it is the collaborative actions of sharing and creating that help to form this safe space, in which dialogues can take place free from any hierarchies of intellect, perceived or otherwise.

6.2 Design

6.2.1 Aims and Objectives

As with any other science communication initiative, there are two main questions that you need to address at the start of planning a collaborative poetry workshop: what are your aims and objectives, and who is your audience [11]? In answering these questions, an aim can be considered as 'what you want to achieve', while an objective should be thought of as 'the action(s) that you will take to realise an aim'. Each objective should be tied to a specific aim, and should also be SMART, i.e. Specific, Measurable, Achievable, Realistic, and Time-bound.

Imagine that there has been a recent policy decision about the use of genetically modified (GM) crops in your region. You want to work with local landowners to better understand their needs, and how their experiences and knowledge of that region could be utilised to ensure an effective and ethical implementation of this new policy. To do this, you want to bring together a selection of agricultural scientists and landowners to discuss the policy and its potential implications. The landowners have been distrustful of agricultural scientists in the past because of an unrelated, non-GM controversy, and you have identified collaborative poetry workshops as a potential method for encouraging dialogue between these two publics.

In this example, you might have the following aim:

To investigate how poetry workshops can be used to encourage dialogue between agricultural scientists and local landowners in relation to GM crops.

Which you could achieve by meeting the following SMART objectives:

- Run three poetry workshops between a group of local landowners and agricultural scientists, generating a series of poems about their attitudes towards GM crops.
- Analyse these poems using poetic content analysis to identify the potential challenges and opportunities for the implementation of related policy.
- Run a fourth workshop with the participants to report findings and confirm the validity of the poetic content analysis.
- Gather feedback from the participants to see how their attitudes change throughout these workshops.

The above example demonstrates that it is difficult to define your aims and objectives without also considering your audience. As discussed in Chap. 2, there is no such thing as a 'general' public, and similarly despite me splitting the global population into 'scientists' and 'non-scientists', these are each multidimensional communities, and are unlikely to be the defining identity for any one such individual.

In the example that was given above the two audiences were more specifically identified as local landowners and agricultural scientists. Each of these audiences represent only one public to which their members belong, but it is this identity that is most important for achieving the stated aim. In this specific example, one might also consider inviting policymakers to the proposed workshops but doing so will undoubtably influence the interactions between these (now three) publics. As such, the involvement of additional publics should largely be avoided, unless they are intrinsic to the aim of the initiative.

Care should also be taken that any one public does not feel undermined or threatened by the presence of another, and that all participants are aware in advance of the other publics that they will be interacting with. It might also be the case that an individual identifies as a member of multiple participating publics (e.g. landowner *and* agricultural scientist). If this occurs, then the capacity in which they have initially been invited to attend should be made clear to the participant, without denying them the opportunity to represent multiple publics where appropriate.

The rationale behind developing your aims and objectives might not be driven by any specific research grant or funding opportunity. For example, you might wish to use a poetry workshop to explore potential collaborations between a research institute and a local community group, or to better understand the areas of scientific research that are important to a set of policymakers. However, from a purely logistical point of view (see Sect. 6.3), it might be necessary to secure some funds to enable the poetry workshops to be run in an inclusive manner.

> **Exercise 6.1: outline your objectives**
> Think about an aim that you would like to achieve through the use of poetry workshops between at least two publics, and why it is that you want to engage this audience. Once you have identified an aim, make a list of 3-5 SMART objectives that will enable you to achieve it, thinking carefully about why poetry workshops are the most appropriate format for doing so.

6.2.2   *Working with the Audience*

Having identified your audience as part of your aims and objectives, how do you go about recruiting them to take part in the workshops? Ideally you will have a point of contact with at least one of the publics that you want to work with, but if not then ask the knowledge exchange team (or equivalent) at your research institute to see if they have any links. If you have no 'warm' introduction, then identify a webpage / Facebook Group / newsletter for the public in question and determine the protocol for contacting them. Reaching out to a representative from each of the publics that you want to work with at the developmental stage of the workshop helps to grant agency and build trust [13]. Furthermore, asking them for feedback on your aims and objectives can help to identify any potential challenges that you might not have been aware of (e.g. cultural sensitives or access requirements), including any pre-existing relationships or tensions with other publics. As will be discussed in Sect. 6.2.3, engaging with representatives of the various publics at this early developmental stage will also likely highlight additional ethical concerns that may need to be addressed.

You should also consider where the workshop(s) is going to take place. Universities and research institutes often represent a physical and psychological barrier for many non-scientific audiences, which only serves to further entrench hierarchies of intellect [14]. Trusted representatives from the different publics that you are working with will be able to provide suggestions, although in some instances a compromise might need to be reached in terms of a 'neutral' venue. Alternatively, if the poetry workshops are to take place over several sessions, then they could potentially alternate between different venues. In some instances, it might also be appropriate to run the workshops in a digital environment (e.g. via Zoom), but in doing so be sure to make allowances for the digital literacy and access for all your participants (see Sect. 6.3).

In terms of recruiting participants for these poetry workshops, it is again advisable to work with the trusted representatives from those publics that you have identified, as they will be able to recommend the best method of advertisement. In some instances, this might involve attending a regular meeting or discussion group, while for other publics mailing lists or social media might be a more effective method of recruitment. Similarly, some publics may require you to build trust with them over longer periods than others, especially those which may be hesitant of engaging with either science, poetry, or the aims and other publics that you are suggesting for your initiative.

Exercise 6.2: engage with your audience

Having identified the audience for your poetry workshops in Exercise 6.1, how will you begin to recruit individuals to participate? Find a trusted member for each of the publics you plan on working with and get in contact with them, outlining your aim and objectives. Do they have any advice for running the poetry workshops or recruiting participants? Are there any ethical considerations that need to be considered when working with this public (or this combination of publics) in particular?

6.2.3 Acting Ethically

As with any science communication initiative, you need to consider the ethical implications of your work. Depending on your research institution or organisation you might be required to receive ethical approval from an independent board of reviewers before you can deliver your initiative, in which case there will be a specific process for you to follow. However, in any case there are a number of ethical considerations that you need to allow for (see Fig. 6.1):

Cultural Awareness. To what extent have you considered the beliefs, needs, and experiences of your publics?

Positionality. What is your relationship to these publics? How do your own beliefs, needs, and experiences potentially bias the ways in which you will interact with them? Will you be participating in the workshops as a member of one of these publics, or remaining as an independent facilitator throughout?

Informed Consent. Is every participant aware in advance of what is expected of them in these workshop sessions? Have you informed them of how any of the outputs may be used, and are all participants aware of the associated risks and benefits of participating in these workshops?

Risks. What are the risks for each of the various publics that are involved, including you as workshop facilitator? To what extent can you ameliorate these risks, and are any of them insurmountable? Remember to consider risks that extend beyond the workshops themselves; for example, participation might be perceived negatively by other non-participating publics or individuals.

Benefits. What are the expectations for all of the publics that are involved? How will participating in these workshops benefit these publics in addition to achieving your aims and objectives for the initiative?

The manner in which these ethical considerations are presented to (and discussed with) your participants will depend on the aim and nature of the initiative. For example, if you plan on analysing the poems that are created in these workshops as part of a publishable research output, then it is advisable to adopt a very formal approach and to present each participant with a participant information sheet and a consent form for them to sign (such an approach is also highly likely to be a requirement of any institutional ethical approval and/or publishing agreement).

If the poetry workshops are instead being run as a dialogue that is not directly linked to research, then it might be more appropriate to have an informal discussion with all of the workshop participants, and to get verbal consent. In any instance, it is your responsibility as workshop facilitator to ensure that *all* participants are informed of the ethical implications of their participation, and that you have put in place the appropriate measures to deal with any risks.

From my personal experience of running these workshops, the two biggest risks to consider are the potential for emotional distress and the revelation of safeguarding information. Writing and sharing poetry can lead to an outpouring of (often unexpected) emotions, and so you need to make sure that you have a set of contingencies to deal with this. Oftentimes it is not the person who is sharing a poem, but other people in the room (including the workshop facilitator) for whom the poem might act as a trigger. As part of establishing a safe space in these workshops (see Sect. 6.2.4), you should make it clear to participants that they can leave at any time, and it is also advisable to set aside a quiet place for reflection and to have someone available to provide emotional support if needed. With regards to safeguarding, writing and sharing poetry in intimate environments can sometimes reveal information about

Fig. 6.1 Ethical
considerations for
developing a poetry
workshop

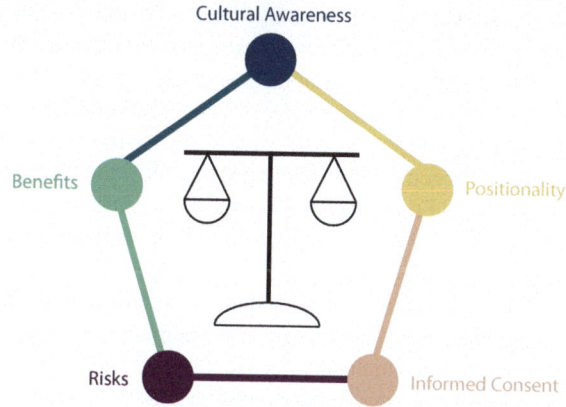

threats to an individual's wellbeing. Be sure to have a system in place to allow for the appropriate reporting of any such concerns that may arise.

Finally, when you are working with potentially vulnerable audiences it might be necessary to get informed consent from a guardian or confidant. If this is the case, then make sure that all parties are fully aware of the correct procedures for obtaining this consent. You might also consider inviting the guardian or confidant to attend the workshop as a participant or observer, if it is appropriate to do so, and providing that the other participants are made aware of this in advance.

Exercise 6.3: consider your ethics

To achieve the aim and objectives that you identified in Exercise 6.1, how will you account for the different ethical considerations shown in Fig. 6.1? Consider the different cultural sensitives of your publics and your position to them as a workshop facilitator (and/or participant). What are the various risks and benefits to all individuals, and how will you ensure informed (rather than assumed) consent for both the workshop and any potential outputs? Having worked your way through each of these considerations, discuss them with the trusted representatives that you identified in Exercise 6.2, and see if there are any additional steps that are needed to safeguard the wellbeing of everyone that will be present at the workshops.

6.2.4   Workshop Plans

Having identified your aims and objectives, contacted representatives from your various publics, and considered your ethics, the next stage is to plan what these poetry workshops will actually involve. Working with your trusted representatives, you should first identify any constraints and then use these to shape a framework for your workshop. For example, you might be limited to a certain length of session, number of participants, or size of venue. In building your workshop plan you should also think about the type of exercise that will be best suited to accomplishing your aims and objectives with these publics. Will the workshop have a reading or writing focus? Should there be an opportunity for participants to share their poems? Will the workshop be a one-off or part of a series? Is there a requirement for a certain set of outputs?

Before I present some examples of different workshop exercises, we need to first consider how to work with participants to establish a safe space. Creating an environment in which all voices feel valued and respected is essential for establishing meaningful dialogue, and for building trust both with and amongst the various publics with whom you will be working.

At the start of any workshop, you should co-create a set of criteria or ground rules that all participants agree to abide to, and which can be referred to as needed. These criteria will depend on the participants in each workshop and can be discussed in advance with trusted representatives from the participating publics; they can also be general or specific, depending on the situation. For example, in addition to agreeing that no one's poetry or opinions should be mocked, it might also be decided that anytime someone wants to speak they should first put their hand up, or that only the person who is holding a designated item (e.g. book, ball, conch shell) can provide feedback at any given time. As discussed in Sect. 6.2.3, you should also outline any safeguarding protocols, or highlight the options that are available to participants if they want to take a break from proceedings because of emotional distress (or for any other reason). Co-creating these criteria also helps to grant agency to the participants in the process. As these are being established is also a good time for the facilitator to ensure that they have the informed consent of all participants.

I will now present four exercises that I have used across a series of different poetry workshops, following which I will discuss what a workshop plan might look like for two different workshop scenarios. These exercises have been chosen in reference to the RAW (Reading, Analysing, Writing) approach to poetry that I outlined in Chap. 2. Feel free to adopt, adapt, and adjust these to meet your own needs, and those of your participants.

Workshop Exercise 1—Share a poem (Reading)

Prior to the workshop inform each of the participants to bring along a poem that they enjoy. This can be on a topic that is related to the aim of the workshop, but it doesn't have to be. For the exercise itself either ask the participants to swap their poem with someone from another public and give everyone time to read this new poem, or ask for volunteers to read their poems out loud to the other participants.

Working with trusted representatives from each of the involved publics beforehand will help you to better understand the form that this exercise should take (i.e. if people should be encouraged to perform their chosen poem, or just be given the time to read another participant's poem in their own time), account for the literacy needs and experiences of your participants, and communicate the requirement of bringing a poem in the most appropriate manner. If it is evident in advance (or becomes apparent during the workshop) that some participants might struggle with reading any of the shared poems, then one option is for the facilitator to read a selection of the poems to all of the participants instead.

Workshop Exercise 2—Discuss a poem (Analysing)

In this exercise participants are paired up with individuals from a different public and asked to 'analyse' a poem. As discussed in Chap. 2, this analysis should focus on what the participants liked or disliked about the poem, and why this was the case. In preparing poems for this analysis, you should choose poems that are related to your aims for the workshop.

For example, if you want to establish dialogue between town planners and ecologists in relation to the environmental impact of a new shopping centre then poems that are related to conservation and/or the built environment would work well (e.g. 'Conservation Status' by Penny Boxall or 'The Truth about Small Towns' by David Baker). This exercise works especially well as a follow on from Workshop Exercise 1, asking participants to work in pairs or small groups and to discuss why they chose to bring their selected poems to the workshop.

In establishing the criteria for a safe space (see above), you might also want to discuss what is appropriate in terms of offering feedback in relation to the poems that the other participants have shared. You might then go on to use the methodology demonstrated in Fig. 2.1 (Chap. 2) to encourage the participants to find out some additional information about the poet and the topic discussed and how this reframes their analyses for each of the poems.

Workshop Exercise 3—List poems (Writing)

This exercise works very well as a warm-up exercise, or to help participants identify words and phrases to use in the development of future poems. Give the participants a set amount of time (normally 60–90 s) to write down a list of everything that they associate with a specified word or phrase. Stress that the list can include physical objects, but also smells, sounds, thoughts, memories, and anything else that pops into their head. This exercise works very well if you repeat it for two or three topics, the first of which could be unrelated to your aims and objectives, but something that your publics might have in common (e.g. a geographical location).

For example, if you were working with a group of local policymakers and a selection of biochemists to explore vaccine hesitancy in the Northwest of England, then your prompts for the list poems might be 'The Northwest', followed by 'Policy' and 'Vaccines'. Inviting participants to share their poems at each stage is an effectual way of both exploring commonalities and encouraging dialogue.

Workshop Exercise 4—Cut and Paste (Writing)

Give every workshop participant a piece of A4 paper and ask them to fold it in half, lengthwise. Prior to the session you should have identified 5–10 words, at least half of which should be related to the aim of the initiative, with the remainder chosen to be either completely random or else linked to the participating publics in some other way. Read one of the words and ask the participants to write this down on the left-hand side of their fold, followed by the word 'is' (or 'are', depending on the word). On the right-hand side of the fold ask them to complete the sentence, with the first phrase that comes into their head. For example, if the word is 'School' then they might write 'School is | a place to learn.' Give them about a minute to write this down, and then ask them to draw a horizontal line under their complete sentence, before repeating the exercise for every word in your list.

Once all of the sentences have been completed, ask the participants to tear the piece of paper down the folded vertical line, and to then make two separate piles of paper. One of the piles should contain the words from your list (followed by 'is/are'), and the other should contain the second half of the completed sentences; the participants can tear along the horizontal lines they drew as a guide. Once they have done this, ask the participants to arrange the list of words (i.e. from the first pile) in a vertical list, and then to draw (without looking) one piece of paper from the shuffled pile of completed sentences (i.e. the second pile), and place it after the first word, repeating this process until a new set of sentences are formed.

If any of the combinations is the same as one of the original (pre-torn) sentences, then ask the participants to mix it up with another one. Give the participants a few minutes to read their new sentences, and then ask for volunteers to share their favourite / funniest / most surprising combinations. This process is shown diagrammatically in Fig. 6.2.

This is a really useful exercise to help people move away from clichés and to start re-examining their relationship with certain words and ideas. A good follow-on exercise is to ask participants to pick one or two of their favourite lines and to use this as the basis for writing a poem about this subject, either in free-verse or via a poetic structure that you have introduced to them (see Chaps. 2 and 3 for some suggestions).

The following two workshop plans have been put together to show what a poetry workshop might look like in practice, depending on your aims, objectives, and publics.

Example Workshop Plan 1

Aim: To investigate how poetry workshops can be used to encourage dialogue between agricultural scientists and local landowners in relation to GM crops.

Objectives:

1. Run two poetry workshops between a group of local landowners and agricultural scientists.
2. At the end of the first poetry workshop the participants should be working towards creating a poem about GM crops.
3. In the second poetry workshop participants will be invited to share their poems with the other contributors.

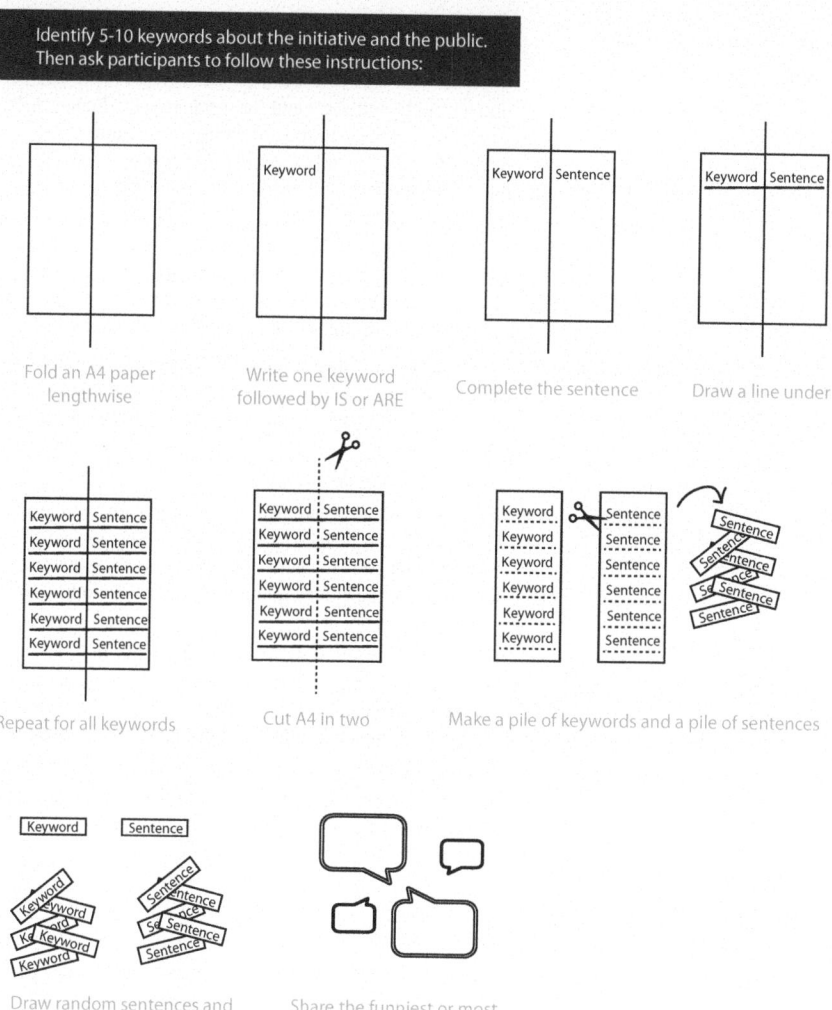

Fig. 6.2 The cut and paste workshop exercise

Publics: Five local landowners, and five agricultural scientists from a nearby university.

Constraints: The landowners are only able to meet after 18:00 on a weeknight and would prefer to meet in a local community centre rather than a university. Having spoken to one of the landowners in advance of the session there is a general mistrust of university researchers by this group as being overly academic and not 'grounded in reality'.

Table 6.1 Workshop Plan 1

Timings (min)	Activity	Notes
0–15	Introduction	Introduce the initiative and the workshop facilitator. Ask everyone to briefly introduce themselves (name plus favourite local amenity), then co-create a set of guidelines for a safe space. Check informed consent
15–25	Share poems	Hand out copies of the three poems that have been chosen for this exercise and ask the participants to read them in their heads. The facilitator should then read them out loud to everyone. These poems have purposefully been picked to showcase examples of free verse and narrative poetry
25–45	Discuss poems	Split participants into pairs (one landowner/one scientist) and ask them to discuss the poems. Which did they like/dislike and why? Potentially invite pairs to share their observations
45–60	Break	Refreshments to be provided
60–70	Cut and paste	Run the 'Cut and Paste' exercise with these 8 words: crops, land, soil, waste, recycle, natural, science, artificial. Invite participants to share their favourite combinations and then ask each participant to pick one line to further develop into a poem
70–85	Poem development	Starting from each of their chosen lines, ask the participants to develop these ideas further into a poem, encouraging them to write in free verse. The facilitator should be on hand to help with any questions or assistance
85–90	Finish	Remind participants to finish their poems for next week, when they will be sharing them with the rest of the group. Reiterate that there is no pressure if they are unable/unwilling to do so

Format: The plan shown in Table 6.1 is for a 90 min session to take place on a weeknight in a local community centre. It is the first of two workshops to take place.

Example Workshop Plan 2

Aim: to explore opportunities for future collaborations with a community radio station and a group of electrical engineers who are researching the use of technologies for healthy ageing.

Objectives:

1. Run three poetry workshops to discuss perceptions of healthy ageing.
2. Use the poems created in these workshops to explore opportunities for future collaborations.

Publics: Ten volunteers from a community radio station and three electrical engineers, all of whom identify as early career researchers.

Table 6.2 Workshop plan 2

Timings (min)	Activity	Notes
0–5	Introduction	Remind participants of their co-created code of conduct and that the workshop will be recorded for the radio. Check informed consent
5–15	List poems	Write three list poems on the following topics: 'This Room', 'Technology', and 'Ageing'. Invite participants to share at least one of their list poems and encourage discussion
15–20	Teach nonets	Teach the participants how to write a nonet, including some examples
20–35	Write nonets	Ask the participants to write a nonet for either 'Ageing' or 'Technology'. The facilitator should be on hand to help with any questions or assistance
35–55	Share poems	Invite the participants to share their poems, and for the other participants to provide feedback. Remind the participants that they are not critiquing the aesthetic quality of the piece, but rather discussing interesting observations and potential synergies
50–55	Q&A	An opportunity for any questions or answers between participants about what has been discussed and shared so far
55–60	Finish	A wrap-up of the session, and a reminder that next week will be the final workshop. Inform the participants that in next week's workshops they will be writing some more poetry that explores some of the topics raised in the workshops thus far

Constraints: The community radio volunteers would like to visit the university and lab space of the electrical engineers as part of these workshops and would prefer for these sessions to take place on a Thursday from 10:00–11:00. All participants are very keen to be involved in this initiative but have indicated that they do not have any additional time to devote to activities outside of the allocated workshops. The workshops will be recorded and later edited together as part of a pre-recorded show for the community radio station.

Format: The plan shown in Table 6.2 is for a 60 min session to take place in one of the laboratories where the electrical engineers test their technologies for healthy ageing. It is the second of three sessions, the first of which was mostly spent introducing the participants to one another and establishing a code of conduct.

After drafting a workshop plan for your initiative, you should then work with trusted representatives from the participating publics to ensure that what you are planning on doing is suitable for the intended audience. You might also consider running a pilot workshop to check timings and address any potential challenges or risks.

Exercise 6.4: develop a programme

Create a programme for a poetry workshop that is designed to achieve the aim you established in Exercise 6.1. Begin by using one of the workshop plans presented in Tables 6.1 and 6.2 and adapt it to both your aim and your intended publics. Remember to consider any constraints that you might have identified in Exercise 6.2 with your trusted representatives. Once you have drafted a workshop programme, share it with these same representatives and ask them for their feedback in terms of the needs and experiences of their publics.

6.3   Delivery

In developing your science and poetry initiative, you should have considered all of the factors discussed in Sect. 6.2, engaged members of the intended publics for their feedback, and perhaps even piloted or trialled your workshop. In terms of delivering a workshop such as that described in Table 6.1 or Table 6.2 there are also several factors that need to be considered. While no two workshops will ever be the same, thinking through the workshop logistics, the role of the facilitator, and the steps that you can take to encourage engagement will help you meet your aims and objectives, with all participants benefitting from the process.

6.3.1   Workshop Logistics

In the developmental stage of your poetry workshop, you should already have determined a venue (or set of venues) that are appropriate for your chosen publics. You should also have worked with representatives from these publics to pick a time of day that is suitable and achievable. When it comes to delivering the workshop there are broadly two sets of logistics to consider: what do you need to do to run the session and how can you make it as accessible and inclusive as possible?

In terms of running the session, the logistics that you need to consider involve making sure that you have all of the materials that are needed for the planned activities and exercises. For example, if you plan on showing a video or playing audio do you have all of the correct equipment, and will the venue enable you use this in the way that you would like? From previous experience, relying on the provision of either specific hardware or a strong Internet connection is not always ideal, and so where possible if you know that there will be a digital component to an in-person workshop I would advise downloading everything in advance onto your own laptop and taking along a set of speakers. If your research institute has some available, you might even consider taking a portable projector.

You should also speak to the venue in advance about any fire or health and safety regulations, and you might be required to fill in a risk assessment by either the venue or your research institute in advance of the workshop. It is always worth checking with your own research institute with regards to the correct protocol for risk assessments and also their requirements for public liability insurance (or equivalent), in the unlikely event that someone were to hurt or injure themselves as a result of attending one of these workshops. Finally, wherever possible you should take along *all* of the equipment that you think your participants will need, including pens and paper. This guarantees their availability, but also if you are working with (for example) a small, volunteer-led community organisation then the provision of such materials can actually be a not-insignificant drain on their resources.

Making the workshops as accessible and inclusive as possible for your intended publics will help to create a safe space for engagement (see Sect. 6.2), but it will also diversify and increase the number of people who can attend the workshops in the first instance. Working with trusted representatives from the publics in advance will help to identify opportunities for inclusion; for example, picking a venue that is accessible for a range of mobilities, choosing times that allow for potential caring responsibilities, and being aware of certain themes or subjects which may act as a trigger for some participants.

In terms of setting up and managing the workshop environment itself, make sure that you are aware of the needs of your participants (including any dietary requirements if you are providing refreshments). You should always aim to finish on time (if not a little early), as your participants might have other appointments to attend; building some contingency time into your workshop plan will help you to achieve this. If you know that some of the participants are not confident in participating in your chosen language for the workshop, or that Deaf people might be in attendance, then you could also arrange for a translator or a signer to be present.

If you are running a poetry workshop in a digital environment, then what have you done to promote inclusion? For example, are you relying on software, highspeed Internet connections, or a degree of digital literacy that might be exclusionary for certain publics? Working with members of your intended publics in the development stage can provide creative solutions for some of these logistical barriers. For example, it might be that multiple participants can share computers, participate via other media (e.g. WhatsApp), or even that the workshop could be run in a hybrid format with some in-person and some virtual participation.

Finally, you need to consider how your participants will be compensated. While not all initiatives may have a budget attached to them that allows for honorariums for all participants, every effort should be made to at least compensate people for their transport costs. Paying people for their participation enables those who would not otherwise be able to participate to take part, while diverse incentives improve participation diversity [15]. If your poetry workshop does not have the budget for such compensations, then instead think about how you might work with your intended publics to maximise their inclusion. For example, are there already events that are taking place within and across these publics, of which your poetry workshop could form a part?

Exercise 6.5: consider your logistics

Take the workshop plan that you developed in Exercise 6.4 and make a list of all of the logistical requirements to deliver this. Consider both what is needed for the delivery of the exercises and what steps you should take to make the workshop as inclusive as possible for your intended publics. Once you have made these lists, show them to the trusted representatives that you have been working with and ask them for their advice and input. If appropriate you can use this list to start to populate a risk assessment form, which you can then show to someone from the legal team at your research institution alongside a request for their policy on running such initiatives in terms of public liability insurance etc.

6.3.2 The Facilitator

The role of the facilitator is something that is often overlooked and yet is one which is fundamentally important to the successful development and delivery of any poetry workshop. In addition to providing a bridge between the various publics, the facilitator must also work with their needs and experiences to foster a creative and inclusive environment. As can be seen from the workshop plans shown in Tables 6.1 and 6.2, the facilitator should also be comfortable in helping participants to engage with poetry, which may also include a familiarity with a wide range of poetic forms, structures, and workshop exercises. They are also likely to need a basic (or more detailed) understanding of some of the scientific topics that are being discussed.

There are many benefits to facilitators freeing themselves from the traditional role of the 'expert', and to instead concentrate on acting as knowledge builders with the rest of the group [16]. Indeed, given that these poetry workshops aim to level hierarchies of intellect between participants, such an approach might seem like the logical method of facilitation. However, from my own experiences it is evident that the facilitator's role in these poetry workshops requires a slightly more removed approach, as they are expected to mediate the different thoughts and opinions within the group, alongside organising the logistics of the sessions. The social learning and development of workshop participants has also been shown to be extended by the support of someone who has a better understanding or higher initial ability level with regards to workshop tasks. The Soviet psychologist Lev Vygotsky referred to this as the 'More Knowledgeable Other' (or MKO) [17] and having a facilitator whom the participants can rely on to assist with their reading, analysing, and writing of poetry is essential for these poetry workshops to succeed in developing dialogue. I would also advise against facilitators participating in the workshops as members of any contributing publics, as doing so can potentially erode trust amongst the other participants.

Given the multiple tasks that a facilitator has to accomplish in these poetry workshops (from organising equipment and timekeeping to writing assistance and conflict resolution), it might be that multiple facilitators are required. Having multiple facilitators can help to accomplish these tasks more efficiently, but for this to be most effective all participants should be made aware of the facilitators and their roles, ideally in advance of the workshop itself. Even if you are the sole facilitator in one of these workshops, I recommend identifying informal 'assistants' from across the publics to help the workshop run as planned and in a manner that is conducive to fostering a safe space and meaningful dialogue. This might include asking for volunteers to act as timekeepers or rapporteurs, or even inviting those trusted representatives who helped in the development of the workshop to ensure that every member of their associated public has the opportunity to be heard.

Finally, as discussed in Sect. 6.2.3 the positionality of the facilitator(s) is vital. A facilitator might not be participating as a public, but they might belong to one (or several) of those that are attending the workshop. Any potential biases should be addressed beforehand; for example, by inviting an additional facilitator to help run the workshop. As well as any conscious biases that a facilitator may possesses, the issue of unconscious bias [18] should also be considered. Unconscious (or implicit) biases are unsupported judgements in favour of (or against) an individual or group as compared to another. One of the ways in which to address these implicit biases is by taking the Implicit Association Test [19], as doing so helps facilitators to become consciously aware of some of the unconscious associations they may possess, and the steps that are needed to address them [20].

Exercise 6.6: reflect on your role

Look at the workshop plan that you developed in Exercise 6.4, alongside the list of logistical requirements that you identified in Exercise 6.5. Do you possess all of the technical abilities to be able to deliver this workshop on your own, and are you able to do so in a way that maximises the benefits for all participants while meeting your aim and objectives? If not, then identify some additional facilitators to help you to deliver various aspects of the workshop. What are the relationships and attitudes of all facilitators with respect to the participating publics? Are there any biases that need to be addressed, and what steps can you take to mitigate these?

6.3.3 Encouraging Engagement

As with any participatory workshop, one of the biggest barriers to success is a lack of engagement. This is particularly true for poetry workshops, where (as discussed in Chap. 2) people might have been discouraged from engaging with poetry or made to feel that in some way it is not 'for them'. A successful poetry workshop is one that meets the aims and objectives of the initiative, but it is also one that benefits all of the participants, and which enables them to be fully engaged. Figure 6.3 presents a handy mnemonic for how to encourage this engagement and help to ensure that neither the participants nor the facilitators are left in a HUFF.

Be Humble. The level of expertise that is demonstrated by a facilitator as the poetic MKO can be difficult to get right. Participants require someone who is able to provide useful feedback and build their confidence in engaging through poetry, without necessarily being so 'brilliant' in the creation or sharing of their own poems as to be disheartening. The extent to which the poetic capabilities of the facilitators are 'showcased' will very much depend on the workshop and the participants; listen to what they want rather than assume what they need. Similarly, don't allow your own needs and experiences to limit your interactions. For example, if you only speak English, but this is not true for the workshop participants, then why limit them to only working with English-language poems? You should never be afraid to be led by the participants, especially if it makes you feel intellectually uncomfortable.

Be Unaesthetic. It is important that any workshop which involves writing (or performing) poetry does not turn into a talent show, in which participants are judged for the 'quality' of what they have created. Doing so will only serve to deconstruct the safe space that has been created, re-establishing hierarchies of intellect with regards to who is the 'better' poet. In these poetry workshops, the focus should not be on the aesthetic quality of the final 'product' but should instead focus on the shared experiences of engaging with poetry, and the dialogues that this elicits. Being unaesthetic does not mean that the poetry has to be unappealing, but rather that its creation (and celebration) should not be motivated purely by aesthetic principles.

Be Flexible. Do not demand or expect the participants to contribute to the workshop in any prescribed manner. For example, it might be that some participants do not

Fig. 6.3 Encouraging engagement through HUFF

want to take part in an activity because they do not feel comfortable doing so at the start of a workshop. However, they may wish to do so later, and denying them this opportunity, will only serve to erode trust and reinforce hierarchies. Being flexible is also driven by your humbleness. Ask the participants if they are enjoying the activities that you have planned and encourage them to critique your suggestions. Remember that you are not the sole beneficiary of these workshops, and as such you should be willing (and able) to adjust to the evolving needs of your participants.

Be Fair. As discussed in Sect. 6.2.4, in establishing a safe space you should be developing an environment in which all voices can be heard. However, this is not limited to the start of any workshop, and in addition to reminding participants of the criteria or ground rules that they have co-created you should also be actively monitoring participant engagement. This is not to say that you should 'force' all participants to share their poetry, but rather that you should give every participant the opportunity to contribute to the workshop in a manner that is most appropriate and beneficial to them. For some participants this might involve encouraging them to share their poetry, whereas for others it might be asking for their opinion or feedback, or even offering to relay this to the other participants on their behalf. It is the responsibility of the workshop facilitator(s) to assure this fairness throughout any poetry workshop, which is why multiple facilitators might be required, especially for larger cohorts.

> **Exercise 6.7: encourage engagement**
> Consider the workshop plan that you developed in Exercise 6.4, alongside the participants that you identified in Exercises 6.1 and 6.2. What steps can you take to help encourage engagement, and to ensure that you are humble, unaesthetic, flexible, and fair in your approach? Revisit your reflections from Exercise 6.6; does this change your opinion in terms of the number of facilitators that might be required for the workshop, or even the way in which the workshop plan itself might be structured?

6.4 Evolution

The final, and most often overlooked, aspect of any poetry workshop is its evolution. This involves reflecting, evaluating, and considering the legacy of the workshop, each of which I will now discuss in turn. The ways in which you reflect, evaluate, and develop legacy for your poetry workshops should be thought about in the developmental stage, as doing so will help to ensure that you have all of the tools and information to do so effectively.

6.4.1 *Reflection*

Reflection is a process of learning through experience, and as well as helping to improve future iterations of any poetry workshop, it is an essential part of developing self-awareness, self-identity, and personal growth as a workshop facilitator [21]. It is something that scientists and science communicators often do in an informal manner; however, adopting a more structured and formal approach enables the most benefit to be gained from the process.

There are several models that can be used to help guide and structure a useful reflection; one of these, Gibbs' Reflective Cycle [22] is shown in Fig. 6.4. Gibbs' Reflective Cycle is centred around the concept that reflection takes place after an experience; it provides a framework of cue questions, offering a checklist for learners as they progress through the cycle.

This reflective cycle focuses on learning from experiences by involving feelings, thoughts, and recommendations for future actions. I find Gibbs' Reflective Cycle to be especially useful as it highlights how 'describing' what happened is only a small part of the reflective process. In using this particular cycle to reflect on a poetry workshop, you should answer each of the questions in turn, starting from the 'Description' of what happened. I recommend writing 25–50 words for each part of the cycle, and to do this as soon as possible after each poetry workshop.

The cyclical nature of the reflective model shown in Fig. 6.4 means that you should be using the action plan to help develop your next poetry workshop, even if this is for a new set of aims, objectives, and publics. For example, after reflecting on

Fig. 6.4 Gibbs' Reflective Cycle. Each stage in the cycle also presents a question to guide the workshop facilitator through a process of meaningful reflection

one poetry workshop you might conclude that there was a lack of engagement for some participants because of an inadequate number of workshop facilitators, and so for a future poetry workshop you might plan to have more facilitators. For a worked example of Gibbs' Reflective Cycle see Sect. 6.5.3.

Gibbs' Reflective Cycle is only one of several formal models that you can use to reflect on your poetry workshops. Other models include Kolb's Cycle of Experiential Learning [23], Rolfe's Reflective Model [24], and the Atkins & Murphy Model [25]. Each of these models has their own strengths and weaknesses, and so it is recommended to try several and see which one allows you to make the most insightful and useful reflections for both your practice and your own personal development as a poetry workshop facilitator. A lot of these reflective models are grounded in nursing and/or pedagogical practice, both of which have a long heritage of formalised introspection and reflection. However, despite their origins, these models can (and should) be adopted for use in reflecting on your poetry workshops (and your role as a facilitator) as part of their evolution.

Exercise 6.8: reflect on an initiative

Consider a recent initiative that you have delivered and use Gibbs' Reflective Cycle and the prompts shown in Fig. 6.4 to reflect, starting with a description, and working your way through to an action plan. Aim to write 25–50 words for each stage in the cycle. If you have recently delivered a poetry workshop (such as that developed in the other exercises in this Chapter) then you can use this for your reflection, but if not then just pick another poetry and/or science communication initiative (e.g. workshop, talk, interview) and use this instead to help familiarise yourself with the process.

6.4.2 Evaluation

In terms of evaluating your poetry workshops, there are two main things to consider: to what extent did you meet your aims and objectives, and what was the experience of the participants?

With regards to your aims and objectives, the manner in which these will be evaluated should already have been determined during the planning phase of your poetry workshop. In picking SMART objectives, you should outline exactly what it is that is 'Measurable' about them, and the steps that you will do to take this. For example, if one of your objectives is to 'run three poetry workshops and use this poetry to form the basis for a poetic content analysis', then you will know that you have achieved this objective if you have run three workshops and generated enough poetry to be analysed using the method outlined in Chap. 4. If you have not met your objectives, then reflect on this process (see Sect. 6.4.1) and identify what this means in terms of achieving your overall aims.

In addition to evaluating your poetry workshops in terms of your aims and objectives you should also consider how to capture the experiences of the participants. Gathering this feedback does not need to be overly elaborate, and can be focussed on a few key questions, such as those shown below:

- What was your experience of the workshop?
- To what extent did the workshop match your expectations?
- What would you have liked to have been done differently?
- How did you hear about the workshop in the first instance?
- Do you have any other comments?

When gathering feedback from your participants, make sure that there is a rationale for every question that you ask, and also that you make it clear to the participants exactly how their feedback will be used. Collecting this feedback should be done in an ethical manner (see Sect. 6.2.3) and should also consider the needs and experiences of the participants. For example, in some instances paper forms might be more appropriate than the use of digital feedback tools such as Google Forms.

The reflections of any facilitators (see Sect. 6.4.1) should also be included in your evaluation and contextualised alongside the feedback from the participants to revisit the extent to which you truly met your aims and objectives. As discussed in Sect. 6.2, it is difficult to delineate your aims and objectives from the publics that you are working with, and as such their experiences of the poetry workshop should form a fundamental role in considering the successes of your initiative.

Exercise 6.9: plan your evaluation

Look at the workshop plan that you developed in Exercise 6.4 and the aim and objectives that you established in Exercise 6.1. How will you be able to monitor the extent to which these were achieved? With regards to the participants of your poetry workshop, what questions should you ask them to evaluate their experiences, and what is the most appropriate format that they should take? Discuss this evaluation plan with the trusted representatives that you identified in Exercise 6.2 and use their feedback to help ensure that this plan is both useful and implementable.

6.4.3 Legacy

You've reflected on your poetry workshop and your role as facilitator and evaluated its success against your aims, objectives, and participant experience. So, what next?

In terms of developing the legacy of these poetry workshops, you need to consider three things: the format of the workshop, the outputs from the workshop, and your relationship with the participants.

In terms of the legacy of the workshop's format, you should use your reflection and evaluation to help develop it for future use. If you plan on running a similar workshop again, then this evolution is relatively straightforward. What worked well in the workshop? What needs to be changed? And what might need to be amended depending on the publics that you are working with? Even if you are not planning on using this exact workshop format again, then there will still be many useful reflections that you can use to help improve future poetry workshops. For example, a participant may have recommended a modification to one of your poetry writing exercises, or someone might have shared a poem that was particularly effective at helping to set the scene for collaborative practice. Use your reflections and feedback to constantly evolve your repertoire of poetry exercises and your expertise as a facilitator. Doing so will help to prepare you for almost any eventuality and will give you the confidence to deal with even the most unexpected of events.

With regards to the outputs of the poetry workshops it is likely that these have been specified in the aims and/or objectives for the initiative. However, these workshops often produce unexpected outputs that can be used to help further the original aim(s). For example, it might be that you had planned a poetry workshop that focused mainly on reading and discussing poetry, but that the participants had also decided that they would like to write some of their own poetry as a result. If you plan on using any of the written outputs from these workshops (in any capacity), then make sure that you have the informed consent of the participants to do so, and that you acknowledge them where appropriate.

In addition to any written outputs, poetry workshops can also produce several key learning experiences that warrant sharing. For example, one of your poetry workshops might result in a lasting collaboration between participants or reveal an unconscious bias of a facilitator in the process. If you plan on sharing these experiences then make sure that you do so ethically, anonymising anyone involved unless they have given their informed consent to be named in such a manner.

Finally, I would strongly recommend providing the participants with your contact information so that they can keep you informed of any future learnings, or just keep you updated with their own poetry and/or science experiences. Doing so also creates opportunities for future collaborations on other science communication initiatives. At the very least, you should try to maintain contact with the trusted representatives who you worked with in developing the poetry workshops. Providing them with feedback on the poetry workshops and acknowledging the role that they played in the process will help to further build trust, which will also benefit any future collaborations with you or others.

Exercise 6.10: consider your legacy

If you have worked your way through the other exercises in this chapter and have successfully delivered a poetry workshop, then think about how you will develop its legacy. Have you considered the evolution of the format that you used based on your reflections and feedback? Were there any unexpected outputs that could be used for future initiatives, and how might you share these in an ethical manner? Do you plan on keeping in contact with any of the participants, and are there any other science communication opportunities where you might be able to collaborate?

If you are still to deliver such a workshop, then think about how you might develop your legacy and the steps that you can take to ensure this. Work with your trusted representatives and use their feedback to help improve the effectiveness of your approach.

6.5 Case Study: TLC St Luke's

To demonstrate what the development, delivery, and evolution of the poetry workshops described in this Chapter might look like in practice, I will now describe a series of such workshops that were developed to initiate dialogue between a scientific and a non-scientific audience.

6.5.1 Development

Aims and Objectives

As discussed in Chap. 4, to successfully mitigate against and adopt to the negative effects of environmental change, it is necessary to engage in dialogue with non-scientific audiences. Doing so raises awareness, builds trust and agency, and also leads to the development of innovative solutions and actions for positive change. However, in engaging in such dialogue, it is often the most vulnerable communities that find themselves neglected. Poetry workshops offer a unique way to bring together scientists and non-scientists, from potentially vulnerable publics, to explore their needs and lived experiences with regards to environmental change.

I have previously shown how poetry can be used to discuss the needs and experiences of different audiences, with relation to environmental change [8, 26]. In this case study I am now adapting some of this work to demonstrate how to develop, deliver, and evolve poetry workshops according to the specific steps that have been presented in this Chapter, beginning with the aim and objectives of such an initiative.

Aim: to use poetry to explore the lived experiences of audiences that are likely to be severely affected by environmental change.

Objectives:

1. Run a series of collaborative poetry workshops with a scientific audience and a non-scientific audience that is particularly vulnerable to environmental change.
2. Generate poems as outputs that could be used to analyse lived experiences with regards to environmental change.
3. Raise awareness amongst the scientific audience for the needs of the non-scientific audience with regards to environmental change.

The first two of these objectives are definitely SMART (i.e. Specific, Measurable, Achievable, Realistic, and Timebound). The third objective is perhaps less easy to measure in terms of whether it is has been achieved or not, but we shall return to this in Sect. 6.5.3.

Working with an Audience

TLC St Luke's is a charity based in Ardwick and Longsight, Manchester in the UK. It was established in 1989 to help people living with mental health needs, by providing a safe and welcoming environment, offering support and activity to all participants. Alongside guidance and advice they also have a dedicated Art Project that enables people, including those with mental health needs, to develop their art practice in a safe, comfortable space.

I was introduced to this community by a colleague who had recommended me for one of the groups' regular Thursday afternoon talks. In this initial introduction I was asked to present some of my work and research on the intersections of science and poetry, and when I first turned up to meet the group (which at the time met at the St Luke's Church and Neighborhood Centre) I had a PowerPoint presentation aimed at a mainly academic audience. After being politely informed that there was no projector, and that the audience were used to a slightly more informal format, I ended up reading a selection of poetry and discussing what feelings it invoked in the attendees.

In preparing for this first encounter, I had failed to consider the needs and experiences of the audience; something which could have been avoided if I had thought to contact a representative from TLC St Luke's in advance. Thankfully, the community members seemed to enjoy the interaction, and over the course of several months I was invited to give several more of these Thursday afternoon talks, developing a strong relationship with several of the community members of TLC St Luke's, as well as their support workers and management team.

I knew that this was a group that would bring an important voice to the project described in the above aim and objectives, and which themselves would also likely benefit from the work. As such I was able to approach the management team, collaborating with them to begin to create a series of workshops that looked at how poetry could be used to give voice to their environmental concerns. It was decided that these workshops should be run in the regular Thursday afternoon slots, and as well as writing poetry, that the attendees would be given the opportunity to discuss their fears and expectations of environmental change with scientific researchers, who I would also invite to participate. It was also recommended that the participants each be provided with a notebook to record their poetry in, which could later be transcribed to meet the aim and objectives of the initiative.

In terms of the scientific audience for these workshops, I recruited several researchers from Manchester Metropolitan University who were working on environmental change research. I was upfront in terms of the time commitment and expectancies of the project, but also of the potential benefits. In selecting these researchers, I purposefully chose from a wide range of disciplines and also made the researchers aware in advance that they might be engaged in dialogues that were somewhat broader than their specific field of expertise, but on topics that were still related to environmental change.

The original research [8] from which this case study is drawn actually involved working with two separate non-scientific publics: members of TLC St Luke's and also Borderlands, a charity organisation based in Bristol for people seeking asylum in the UK or who have become a refugee from other countries. However, these two publics presented different challenges and opportunities for engagement, and so for the purpose of this case study I am only considering those poetry workshops that took place at TLC ST Luke's.

Acting Ethically.

This case study is part of a larger research project that received full ethics approval via Manchester Metropolitan University's Academic Ethics Committee. According to Fig. 6.1, the following ethical considerations were considered in the development of these workshops:

Cultural Awareness. There was a danger that the members of TLC St Luke's who were participating in this study might be perceived as having been chosen for exploitative reasons. However, this audience was selected because they represented a community that is underrepresented in terms of environmental change. Prior to the poetry workshops I worked in tandem with representatives from this community to ensure that all participants knew exactly what the study was for, what it would entail, and what their involvement would consist of. All participants were also made fully aware of the study, and that anyone could take part in the activities without writing any poetry or having any of their poetry recorded. As such everyone could potentially benefit from these sessions, and the danger of exploitation was significantly reduced.

Positionality. At the time of these workshops, I was an academic working at Manchester Metropolitan University. I therefore needed to consider my role as a colleague to the scientific researchers that I had invited to attend these workshops. Similarly, as discussed above I had also started to develop a relationship with TLC St Luke's, and so I needed to be careful not to take advantage of the position of trust that I had established. During the workshops it was necessary for me to state my role as an impartial facilitator, who was there as neither a scientist nor a non-scientist.

Informed Consent. Before participating in this workshop, all participants were issued with a participation information sheet and a consent form, which they were asked to sign. With regards to the non-scientific audience passive assent was avoided by working with the TLC St Luke's management team to make sure that the research complied with the Human Rights Act, the Mental Capacity Act, and the Equalities Act. All the participants were given sufficient time to read the consent forms. Participants were not coerced into taking part, and the support workers at TLC St Luke's had access to the consent forms prior to the sessions, so that they could explain the aim of the workshops to the participants, if necessary. The forms were also carefully discussed in the sessions, and the participants were given the option of taking them away to study them further if required.

Risks. I worked closely with the support workers to ensure that a correct system was established for safeguarding, in relation to any information that may have been revealed during the sessions or from the subsequent analysis of the poetry. During these sessions, and the subsequent analysis, no such information presented itself. Similarly, there was also a risk to the participants that some of the poems might have acted as an emotional trigger, and this was discussed before each session, with participants invited to step outside whenever they felt the need to do so. During these sessions no such trigger occurred. With regards to the transcription of the poems produced in these workshops, anonymity was preserved by prescribing a pseudonym to all the poetry. Following this measure there was still the risk that some of the poetry might contain identifying narratives, and so during the analysis, any specific or personal narratives that could be identifying were redacted and destroyed without recording them.

Benefits. The benefits to the non-scientific audience were that they would have the opportunity to discuss their needs and experiences about environmental change in a forum specifically designed to allow them to do so. The benefits for the scientific audience were that they would be able to find out more about how environmental change was affecting a potentially vulnerable audience, and to discuss how their subsequent research might help to address these needs. All participants also benefitted from becoming more familiar with their use of poetry as a result of these workshops. Finally, these poetry workshops also benefitted TLC St Luke's by providing content for three of their Thursday afternoon talks.

Workshop Plans

The plan shown in Table 6.3 is for a 60-min session that took place at TLC St Luke's on a Thursday afternoon, during their weekly series of talks which also included a hot meal for all participants (after the workshop). This was the second of three sessions, the first of which was mostly spent introducing the participants to one another, creating a safe space, outlining what the purpose of these workshops were, and writing poems about climate change. At this second session there were three scientists and 15 non-scientists in attendance, although as will be discussed in Sect. 6.5.2, there was some fluidity with regards to who attended each session.

Table 6.3 Workshop plan for environmental change

Timings (min)	Activity	Notes
0–5	Introduction	Check informed consent. Hand out notebooks to participants
5–10	Aerosol discussion	In the last session the non-scientists had asked what was meant by 'aerosol' so a brief explainer to be given by one of the scientists
10–15	List poems about aerosol	Write list poems about aerosol
15–25	Share list poems	Invite participants to share and discuss their list poems. Facilitator to be on hand to help
25–30	Pollution in Manchester	In the last session the non-scientists had asked to find out a bit more about air pollution in Manchester. Facilitator to provide a brief explainer, including print outs of air pollution readings in this region
30–40	Poems about pollution	Ask the participants to write poems about air pollution based on the discussions that have taken place thus far. Facilitator to be on hand to help
40–50	Share poems	Invite participants to share and discuss their poems
50–60	finish	Open Q&A, ask what the focus should be for next week. Collect notebooks

6.5.2 Delivery

Workshop Logistics

In delivering the poetry workshops for this Case Study, what did I do to run the sessions and how did I make them as accessible and inclusive as possible?

In terms of the logistics of running the poetry workshops themselves, one of the main challenges was how to record the poems that were produced in these workshops. As can be seen from Table 6.3, the participants were asked to record their poems in notebooks, which were provided by me and were handed out at the beginning of each session and collected at the end. This decision was made following consultation with the support workers, who thought that this would be the most effective way of recording the poems that were created in the workshops, and that this method would place no additional stress on the participants in terms of having to remember to bring their notebooks with them to each of the sessions.

As discussed in Sect. 6.5.1, these poetry workshops took place on three consecutive Thursday afternoons, and each lasted for approximately 60 min. During the poetry workshop refreshments were provided to all participants in the form of tea, coffee, juice, and biscuits, and after each of the workshops all participants were invited to stay for a hot, cooked meal. This followed TLC St Luke's regular format for their Thursday afternoon series of talks.

With regards to making the event as accessible and inclusive as possible, I worked with the TLC St Luke's management team and support workers to advertise the poetry workshops in advance. We decided that it would be best for the workshops to take place in the foyer of the St Luke's Church and Neighbourhood Centre (where the regular Thursday afternoon talks usually took place). We also discussed how I might engage the participants in my role as a facilitator (see below), and we decided to avoid changing the regular set up of the foyer to avoid any confusion.

There was no financial incentive provided to any of the participants, but a donation was provided to TLC St Luke's to help pay for their staffing costs, the hiring of the space, and the cost of the refreshments (and hot food) that was provided. This donation was costed into the research budget for the larger research project that this case study formed a part of. In this instance not providing any financial incentive to the participants did not reduce the accessibility and diversity of the event, as the workshops took place in the regular Thursday afternoon talks that were run by TLC St Luke's. The scientists that were involved in the poetry workshops were first met by me prior to entering the venue, and in the first session I helped to facilitate an introduction between all of the participants.

The Facilitator

Prior to running these poetry workshops I had run several such workshops in other locations, and I felt confident that I would be able to conduct these workshops without any additional facilitators. My main roles involved creating a safe space, helping participants to transcribe and read their poetry (see below), timekeeping, and ensuring that the workshops were carried out in an ethical and inclusive manner. I also had to remember to bring the notebooks and biscuits to each session, both of which were critical for the effective delivery of the workshops.

As discussed in Sect. 6.5.1, I needed to consider my position as both a colleague to some of the participants (the scientists), and a trusted confidant of the others (the non-scientists). Explicitly stating that I would not be creating my own poetry helped to demonstrate to all participants that I was unbiased in my facilitation, and by not reading any of my own poetry I also avoided the danger of the participants potentially deferring to me before sharing their own work. Furthermore, by visibly helping several of the participants to craft their own poetry I was able to demonstrate my role as an MKO, thereby helping the participants to develop their own confidence in engaging with poetry.

Encouraging Engagement

According to Fig. 6.3, I encouraged engagement by adopting the following approach:

Be Humble. As discussed above I did not write or share any of my own poetry in these workshops, even though initially the temptation to do so had been great. In many of the previous poetry workshops that I had run (prior to those in this case study) I had shared my own work, but on reflection I found that this had led to some hesitancy in the participants' willingness to share their own poems. This proved to be a good decision, as the poems that were created in these workshops were incredibly effective at generating dialogue and helping to develop empathy between the participants.

Be Unaesthetic. After every poem that was shared, I encouraged the participants to show their appreciation with a small round of applause, and this was later automatically taken up without any prompting from me. In truth some of the poems were more aesthetically pleasing in a traditional sense, but the context that was often provided by the participants when they shared their poetry meant that no poem could truly be perceived as 'more valuable' or 'better', as reflected by the appreciation shown after each poem was shared. In hindsight I probably enabled one of the participants to share their poetry more regularly than any of the others, as this participant often wrote provocative and/or insightful poetry that resulted in lots of discussion. However, the participants also seemed to enjoy this, and I was always careful to ensure that everyone was able to share their poetry if they so wished.

Be Flexible. As discussed in Sect. 6.5.1, the number of non-scientific participants for any one of these poetry workshops was somewhat fluid. This is because some participants chose to engage throughout, some took regular intervals (for refreshments, cigarette breaks, etc.), some left early, some arrived late, and some decided to participate at the very end of the session having not been present for any of the former part. Having previously worked with this audience I knew that this was likely to be the case, and I had briefed the scientists to expect this as well. However, by encouraging participants to engage in any way that they wished (even if it was not at all in keeping with the workshop plan) I was able to foster a sense of trust and inclusivity. As will be discussed in Sect. 6.5.3, this flexibility was not always evident, and was something that evolved during the initiative. In some instances I was not able to gather informed consent for the recording of the poems that were written, but this did not in any way detract from the richness of the experience, and indeed such interactions undoubtably inspired the creation of new poems that could be captured and transcribed in an ethical manner.

Be Fair. Some of the participants liked to read all of their poems, some of them were published poets, some of them were illiterate, and some of them did not like to read their work out loud. During the poetry writing part of the workshops I tried to engage with all of the participants on an individual level. Sometimes this involved transcribing the poems for them, sometimes it involved editing their work, sometimes it involved reading on their behalf, and sometimes it involved just listening to what they had to say. By treating each of the participants as individuals with their own needs and experiences I was able to foster a greater sense of engagement than if I had expected them to engage in an identical manner. The size of the workshop and my relationship with the participants made this especially effective, but such an approach can also be utilised for both a larger (have more facilitators) and less familiar (spend more time building trust and listening) audience.

6.5.3 Evolution

Reflection

Using Gibbs' Reflective Cycle (Fig. 6.4) I will now reflect on the poetry workshop that is outlined in Table 6.3, and in particular one event that occurred during this workshop.

Description. During the workshop some members of the TLC St Luke's community arrived late and were a little disruptive in the discussion of aerosols.

Feelings. I was annoyed that these individuals were having a negative effect on the engagement of the other participants. However, I then realised that I was being overly controlling in the ways in which I wanted people to engage with the workshop, which made me feel ashamed.

Evaluation. It was initially bad that these individuals disrupted the workshop, but they actually had some very valuable insights which prompted further discussion.

Analysis. I perceived this situation to be 'bad' because it did not quite fit into the ways in which I wanted participants to engage with the workshop. However, once I got over myself, I realised that these individuals actually wanted to engage, just in a different way, and that I needed to create a space for them to do so.

Conclusion. Eventually these individuals became very involved in the workshop, but in hindsight I should have made them feel more welcome in the first instance and not been so preoccupied with my timekeeping for the session.

Action Plan. For the next workshop I decided to make sure that I allowed space for latecomers to engage with the workshop in a way that was valuable to them. I also planned to be less obsessive with sticking to time with regards to certain activities.

As can be seen from the above reflections, I used this cycle to identify an action plan to help improve future workshops, which I was then able to implement in the third and final workshop for this initiative. This process also helped me to identify failings in my facilitation, which I was able to address to improve the experience for future initiatives as well. Such reflections should ideally take place both after each individual poetry workshop and at the end of any series, to better enable the evolution of both the workshop and your own practice.

Evaluation

To what extent did I meet my aim and objectives, and what was the experience of the participants? The objectives of this initiative, as initially stated in Sect. 6.7.1, were to:

1. Run a series of collaborative poetry workshops with a scientific audience and a non-scientific audience that is particularly vulnerable to environmental change.
2. Generate poems as outputs that could be used to analyse lived experiences with regards to environmental change.

3.   Raise awareness amongst the scientific audience for the needs of the non-scientific audience with regards to environmental change.

The first and second objectives were clearly met, in that a series of collaborative poetry workshops was run (Objective 1), during which poetry was generated for subsequent analysis (Objective 2).

With regards to Objective 3, this was also related to the experiences of the scientific participants, whom I asked to reflect on their experiences through poetry. The following poem (written by one of the scientific participants) is representative of this feedback, and demonstrates the extent to which Objective 3 was achieved:

Climate change is real here and there

We all have knowledge that we should share

Although it may seem a complex affair

Poetry can help us, that I declare!

Some people were talkative, others were shy

Some wore a tracksuit, others a tie

I was out of my comfort zone, I cannot deny

But the stimulating vibe encouraged me to try

For the next time, please save me a seat

Through poetry we can all again meet

I liked exchanging ideas, no need to compete

Experts and non-experts together, climate change will beat!

In meeting these objectives, I also believe that I achieved my aim, which was to use poetry to explore the lived experiences of audiences that are likely to be severely affected by environmental change.

In terms of the experience of the non-scientific participants, the feedback that I received from the support workers and management team at TLC St Luke's was that it had been a valuable experience, in which their members had enjoyed participating. Some of the poetry that was written by the non-scientific participants during these workshops also demonstrated the extent to which they valued the dialogues and conversations that took place there, as is evident from the following poem, written in the third workshop in response to the prompt of 'making a change':

Everyone has change which is a

Part of life you don't know

What change will bring and you should

Educate yourself about climate ask

Questions.

Legacy

As discussed in Sect. 6.4.3, there are three areas to consider with regards to the legacy of this case study: the format of the workshop, the outputs from the workshop, and my relationship with the participants.

In terms of the format of the poetry workshops used in this case study, as can be seen from the reflections discussed above, I learnt some very valuable lessons about being overly restrictive with timings and how people 'should' engage. This was a formative learning experience for me, and it gave me the confidence to be much more relaxed in my use of a workshop plan. I now know that while having a workshop plan is extremely useful, it need only act as a rough guideline, and that the engagement (and enjoyment) of the participants is of much greater importance than making sure I have neatly 'ticked off' all of my timings.

There was significant legacy from the outputs of this workshop, as the poetry was used to perform a poetic inquiry which was subsequently published as part of a research project investigating the use of poetry to give voice to underserved audiences in relation to environmental change [8]. However, if I am being critical of both myself and the case study, then I would say that this did not go far enough, and that actually the outputs should have been used to enact change; for example by helping to inform local policy or raising awareness in the local government for the needs of this specific community.

Finally, there is the legacy of my relationship with the participants to consider. As a result of these poetry workshops I collaborated with the participating scientific researchers on several further science communication projects (several of which involved poetry). I also further strengthened my relationship with TLC St Luke's and have since been invited back several times to read, write, and discuss poetry. These subsequent workshops and discussions have not been related to any research study but have simply arisen from the enjoyment of poetry that I helped to foster with this community. This is probably the most important and successful output, outcome, or legacy that I achieved with the work presented in this case study.

6.6 Summary

In this chapter, I have presented a detailed explanation of how to develop, deliver, and evolve poetry workshops. I have also discussed why these poetry workshops are an effective format for developing dialogues between scientific and non-scientific audiences, and how they can be used to level hierarchies of intellect. By presenting a detailed case study from my own practice, I have also demonstrated what each of these steps looks like, from setting the aims and objectives of such an initiative right through to reflecting on the workshops and using this to help develop a legacy. If you have followed the exercises in this Chapter, then you will also have developed your own poetry workshops and should hopefully now be in a position to deliver and evolve them. The most important aspect for any of these workshops are the participants, and every decision that you make should be done to ensure that they are treated ethically and with respect. Listen carefully to what they need, rather than assuming what they want.

6.7 Suggested Reading

In addition to the case study presented in Sect. 6.5, 'Representing the majority and not the minority: the importance of the individual in communicating climate change' [26] presents a detailed description of how to develop, deliver, and potentially evolve poetry workshops for three distinct community groups. In particular, this research highlights the value of engaging those audiences that have been largely ignored in the development of science communication dialogues, demonstrating how poetry workshops can act as a powerful tool for empowering citizens and enacting change. For those readers who struggle to reflect on their practice, *The Reflective Practice Guide* by Barbara Bassot [27] is an excellent resource, providing a comprehensive review of the literature on reflective practice as well as practical examples to help develop your skills in critical reflection. Finally, *Poetry and Dementia: a Practical Guide* by John Killick [28] presents a range of techniques and exercises for engaging different publics with poetry, including many practical exercises and also a detailed consideration of the ethical issues of running poetry workshops with potentially vulnerable audiences.

6.8 Further Study

The further study section in this Chapter is designed to help you develop, deliver, and evolve your own poetry workshops as an effective medium through which to engender dialogue between scientific and non-scientific audiences.

1. **Reflect on a workshop**. Use the reflective cycle shown in Fig. 6.4 to reflect on a recent workshop in which you were a participant. This can be any workshop, and it need not be related to science and/or poetry. To what extent were your experiences influenced by the facilitator(s) of that workshop, was there anything that you could borrow for your own facilitation, and what did not work particularly well? Reflecting on your experiences as a participant is a useful way to help develop your own practice as a facilitator, and you should try to do so at every opportunity.

2. **Develop an activity**. Taking the poetry workshop exercises presented in Sect. 6.2.4 as inspiration, develop your own activity. Think about whether it will focus on reading, analysing, and/or writing poetry, and what outputs it should generate, if any. What steps can you take to ensure that this activity is

accessible for all of your intended participants, and how might it be adapted for different initiatives?

3.   **Get upskilled**. Referring to Exercise 6.6, consider your role as a facilitator for a poetry workshop. Are there any skills that you would like to improve to help you design, deliver, and evolve poetry workshops more effectively? For example, do you lack confidence in providing feedback on people's poetry, or do you feel that you would benefit from having some additional workshop exercises to draw from? Identify opportunities for improvement (e.g. workshops, seminars, video tutorials) and use these to match your needs. As with the first Further Study exercise in this Chapter, be sure to reflect on your experiences as a learner / participant and how you can use this to improve your own poetry workshops in the future.

References

1. Nisbet MC, Scheufele DA (2009) What's next for science communication? Promising directions and lingering distractions. Am J Bot 96(10):1767–1778. https://doi.org/10.3732/ajb.0900041

2. Lorenz L (2020) Addressing diversity in science communication through citizen social science. J Sci Commun 19(4):A04. https://doi.org/10.22323/2.19040204

3. Van der Sanden MCA, Meijman FJ (2008) Dialogue guides awareness and understanding of science: an essay on different goals of dialogue leading to different science communication approaches. Public Underst Sci 17(1):89–103. https://doi.org/10.1177/0963662506067376

4. Berdahl L, Bourassa M, Bell S et al (2016) Exploring perceptions of credible science among policy stakeholder groups: results of focus group discussions about nuclear energy. Sci Commun 38(3):382–406. https://doi.org/10.1177/1075547016647175

5. Bonney R, Cooper CB, Dickinson J et al (2009) Citizen science: a developing tool for expanding science knowledge and scientific literacy. Bioscience 59(11):977–984. https://doi.org/10.1525/bio.2009.59.11.9

6. Muradova L, Walker H, Colli F (2020) Climate change communication and public engagement in interpersonal deliberative settings: evidence from the Irish citizens' assembly. Clim Policy 20(10):1322–1335. https://doi.org/10.1080/14693062.2020.1777928

7. Einsiedel EF, Eastlick DL (2000) Consensus conferences as deliberative democracy: a communications perspective. Sci Commun 21(4):323–343. https://doi.org/10.1177/1075547000021004001

8. Illingworth S, Jack K (2018) Rhyme and reason-using poetry to talk to underserved audiences about environmental change. Clim Risk Manage 19:120–129. https://doi.org/10.1016/j.crm.2018.01.001

9. Illingworth S (2020) Creative communication–using poetry and games to generate dialogue between scientists and nonscientists. FEBS Lett 594(15):2333–2338. https://doi.org/10.1002/1873-3468.13891

10. Jack K, Illingworth S (2017) 'Saying it without saying it': using poetry as a way to talk about important issues in nursing practice. J Res Nurs 22(6–7):508–519. https://doi.org/10.1177/1744987117715293

11. Illingworth S, Allen G (2020) Effective science communication, 2nd edn. Institute of Physics Publishing, Bristol

12. Watermeyer R, Montgomery C (2018) Public dialogue with science and development for teachers of STEM: linking public dialogue with pedagogic praxis. J Educ Teach 44(1):90–106. https://doi.org/10.1080/02607476.2018.1422621

13. Illingworth S (2018) Delivering effective science communication: advice from a professional science communicator. Semin Cell Dev Biol 70:10–16. https://doi.org/10.1016/j.semcdb.2017. 04.002
14. Mackay SM, Tan EW, Warren DS (2020) Developing a new generation of scientist communicators through effective public outreach. Commun Chem 3(76). https://doi.org/10.1038/s42 004-020-0315-0
15. Hsieh G, Kocielnik R (2016) You get who you pay for: The impact of incentives on participation bias. In: Proceedings of the 19th ACM conference on computer-supported cooperative work & social computing (CSCW '16). Association for Computing Machinery, New York, p 823. https://doi.org/10.1145/2818048.2819936
16. Blake J, Illingworth S (2015) Interactive and interdisciplinary student work: a facilitative methodology to encourage lifelong learning. Widen Participation Lifelong Learn 17(2):108–118. https://doi.org/10.5456/WPLL.17.2SI.107
17. Vygotsky LS (1980) Mind in society: the development of higher psychological processes. Harvard University Press, Cambridge
18. Dee T, Gershenson S (2017) Unconscious bias in the classroom: evidence and opportunities. Google Inc., Mountain View
19. Nosek BA, Bar-Anan Y, Sriram N et al (2014) Understanding and using the brief implicit association test: recommended scoring procedures. PLoS ONE 9(12). https://doi.org/10.1371/ journal.pone.0110938
20. Staats C (2016) Understanding implicit bias: what educators should know. Am Educ 39(4):29–33
21. Wain A (2017) Learning through reflection. Br J Midwifery 25(10):662–666. https://doi.org/ 10.12968/bjom.2017.25.10.662
22. Gibbs G (1988) Learning by doing: a guide to teaching and learning methods. Oxford Polytechnic Further Education Unit, Oxford
23. Kolb DA (2015) Experiential learning: experience as the source of learning and development, 2nd edn. Pearson Education Inc., Upper Saddle River
24. Gary R, Freshwater D, Jasper M (2001) Critical reflection in nursing and the helping professions: a user's guide. Palgrave Macmillan, Basingstoke
25. Atkins S, Murphy K (1995) Reflective practice. Nurs Stand 9(45):31–35
26. Illingworth S, Bell A, Capstick S et al (2018) Representing the majority and not the minority: the importance of the individual in communicating climate change. Geosci Commun 1(1):9–24. https://doi.org/10.5194/gc-1-9-2018
27. Bassot B (2015) The reflective practice guide: an interdisciplinary approach to critical reflection. Routledge, London
28. Killick J (2017) Poetry and dementia: a practical guide. Jessica Kingsley Publishers, London

Chapter 7
Poetic Collaborations

7.1 Introduction

In Chap. 6, I introduced the concept of a poetry workshop and discussed how this could be used as an effective method through which to develop dialogue between scientific and non-scientific publics. Aside from using poetry as a facilitatory medium for such conversations, it is also valuable to explore the potential for collaborations between scientists and poets to diversify science and develop further opportunities for science communication.

There exist many excellent examples of interdisciplinary collaborations between creatives and scientists. From a UK perspective, the Science Ceilidh [1] is an award-winning educational organisation based in Scotland that aims to bring people together with science and traditional music and arts. Another well-established collaboration is Invisible Dust [2], a charity that aims to facilitate dialogue between leading visual artists, creative technologists, and scientists to encourage meaningful responses to climate change and environmental issues. Similarly, the Arts Catalyst [3] is an arts and educational charity that works across art, science, and technology to produce projects that critically engage with our changing world. Finally, Cape Farewell [4] uses art, music, and culture to engage the wider public about climate change and environmental challenges, inviting artists and creatives to join scientific expeditions that have explored the arctic, sustainable island communities, and areas of urban regeneration.

Away from the UK, the Center for Advanced Visual Studies at the Massachusetts Institute of Technology (MIT) has been exploring the role of technology in culture for over half a century [5], while S+T+ARTS (Science, Technology & the Arts) is an initiative of the European Commission to support collaborations between artists, scientists, engineers, and researchers to develop more creative, inclusive, and sustainable technologies [6]. Other notable examples of innovative collaborations include: Catching a Wave, a multi-media installation about sea-level rise designed to elicit action towards more sustainable oceans and coasts [7]; the Community Engagement in Science Through Art (CESTA) summer programme, which has brought together

© The Author(s), under exclusive license to Springer Nature Switzerland AG 2022
S. Illingworth, *Science Communication Through Poetry*,
https://doi.org/10.1007/978-3-030-96829-8_7

chemistry, sculpture, and engineering undergraduate students to design and construct interactive art to engage their local communities with chemistry [8], and Guerrilla Science, who utilise the interface of science and art to connect and engage with publics who do not otherwise take part in science activities [9].

In addition to programme-specific collaborations between artists and creatives, Arts at CERN [10], the SETI Institute Artists in Residence programme [11], and the Antarctic Artists & Writers Program [12], all provide examples of residencies that afford creatives the opportunity to collaborate with scientists to both communicate science to new audiences and also develop new and innovative directions for scientific research and governance.

The successes of these interdisciplinary collaborations are grounded in fostering an environment in which all participants are treated as valued and respected partners. In this Chapter, I will present a manifesto for developing such collaborations between scientists and poets, demonstrating how it can be used to create partnerships that are fair, just, and successful at achieving their aims and objectives. This manifesto has been developed based on my own work and practice, including the many formative learning experiences that have resulted from both leading and participating in such interdisciplinary collaborations. I conclude this Chapter by using use two case studies to demonstrate how scientists and poets can collaborate in practice, framing these partnerships within the context of this manifesto.

7.2 A Manifesto for Poetic Collaborations

A manifesto is a method of communication that specifically addresses an audience and asks them to unite to enact change. Manifestos are often associated with government agendas, and indeed some of the earliest and best-known examples (such as the 'American Declaration of Independence' or the 'Communist Party Manifesto') are grounded in politics, and in particular radical democracy. However, manifestos also have a strong poetic heritage, from the 'Dada Manifesto' of Tristan Tzara in the early twentieth century [13] through to the Hate Socialist Collective's 'Leave the Manifesto Alone: a Manifesto', published in 2009 [14]. Similarly, science has a history of impactful manifestos, perhaps most famously the 'Russell-Einstein Manifesto' of 1955 [15] which warned against the dire consequences of a nuclear war, and more recently the 'Scientific Manifesto Based on the Philosophy of Science of Karl Popper' from the Open Science Repository [16], which is committed to making scientific results freely available for everyone to debate in an open and constructive manner.

Manifestos do not typically adhere to any one specific style or format, but rather they are united by a desire to urge the reader into action, thus making them particularly effective for helping to develop a framework for interdisciplinary collaborations. Another aspect of manifestos is that they tend to be written from an opposing position, implying that there is an 'us' (the people wanting to enact change) and a 'them' (those opposed to it). However, in developing this manifesto for poetic collaborations

I purposefully avoid adopting such a position. Rather, I present a set of recommendations grounded in my own experiences, in the hope that they can encourage future interdisciplinary collaborations between scientists and poets, that are fair, equitable, and beneficial for all involved.

1. Begin at the Start

Collaborations should begin at the start, and not at the end. Asking a poet to write a poem about some recently completed research is not really an interdisciplinary collaboration, it is a commission. By involving all collaborators as early as possible, the project will stand to gain the most benefit from all of the skills and expertise that they possess. For example, if you are thinking of applying for a grant to facilitate a collaboration between a group of scientists and poets, then why not invite the poets to be involved in the grant writing process itself?

Similarly, in constructing your research questions, aims, and objectives, why limit this to only the scientists that will be involved in the collaboration? It is not only scientists that have research or grant-writing expertise that such a project would benefit from. Presenting all collaborators with the opportunity to engage with the process from the very beginning will develop trust and agency in the collaboration and will also give everyone involved more time to learn how to work together as a team. Furthermore, increasing the diversity of contributions from the very start will normally lead to innovative ideas that enrich experiences for all participants.

2. Grant Agency

Do not limit the way that participants want to engage with a collaboration. As discussed in the first point of this manifesto, poets should be encouraged to become involved beyond the writing and performance of any poems. Similarly, if there are scientists who want to write poetry or do something else creative then they should be encouraged to do so. Even if you have initially approached someone to join a collaboration because of a particular skillset that they possess, you should not make their involvement conditionary on them exclusively utilising that specific skillset.

Academia has a nasty habit of sorting people into well-constrained silos, which is in part to blame for the lack of diversity that we often encounter in science and scientific research more broadly. Use your collaborations as a vehicle of change to enable your collaborators to break free from these silos, and to contribute in a way that is also beneficial to them.

3. Reward Involvement

The majority of professional poets work as freelancers, without the security of a full-time position, sick leave, holiday entitlements, or pension contributions. Which is to say that wherever possible you should try to pay any poets (or other creatives) that you are collaborating with. Alternatively, make sure that any benefits are commensurate with the time and effort that is being expended. This might be in the form of professional development, peer-to-peer learning, access to specialist equipment or materials, or the possibility for future payment (for example when helping in the writing of a grant application). Whatever you do though, please do not offer 'exposure' as a form of payment; the poets that you are working with are already well-enough established for you to have been aware of their work in the first instance.

In terms of rewarding scientists for their involvement in the collaboration, the argument for financial reparation is less straightforward. If the scientists that you are working with are in permanent, full-time positions then they might not require financial compensation because their involvement in the collaboration could fit within the boundaries of their paid employment. However, this is not always the case, especially for those scientists working outside of their scientific remit, and definitely for those scientists who are employed on temporary or fractional contracts. In such instances you should be upfront with the scientists in terms of any potential benefits and rewards. If financial compensation is not sought and/or not possible, then discuss the opportunity for them to participate so that it is meaningfully beneficial to them in other ways, as discussed in point two of this manifesto.

4. Be Humble

Being precious about your work and ideas is not a good way to elicit meaningful collaboration. You should be open to suggestions and feedback for how to improve both the collaboration and your role in it. Sometimes this feedback might be unwarranted or hard to hear, especially if it is related to something that you have tried to account for. However, creating a space in which you encourage constructive criticisms and supportive reflections will help to foster trust across the collaboration. There is of course a certain amount of pragmatism that is needed when incorporating such feedback, as there might be deadlines or specific deliverables to adhere to. Being humble also means that you should take responsibility for the collaboration, that you should own any mistakes, and that you should commit to learning from these for the benefit of both the collaboration and your own development.

5. Encourage Evolution

Interdisciplinary collaborations need not stop when they have met their aims and objectives. Rather, they should be encouraged to continuously evolve both within and beyond the framework of their initiation. Being humble and granting agency will help to nurture an environment in which collaborators are free to experiment and encouraged to take the collaboration in new and innovative directions. Most interdisciplinary collaborations are 'doomed to initially succeed'. That is to say, such collaborations often have enough energy and goodwill behind them to achieve their initial aims but lose momentum shortly afterwards. Those interdisciplinary collaborations that go on to develop a lasting legacy are usually those that are free to evolve, which adapt to the needs and interests of the collaborators, and which are not overly reliant on any one individual.

6. Listen

The final, and most important point of this manifesto for science and poetry collaborations is to listen. Listen to your collaborators and in turn encourage them to listen

Fig. 7.1 A six-point
manifesto for science and
poetry collaborations

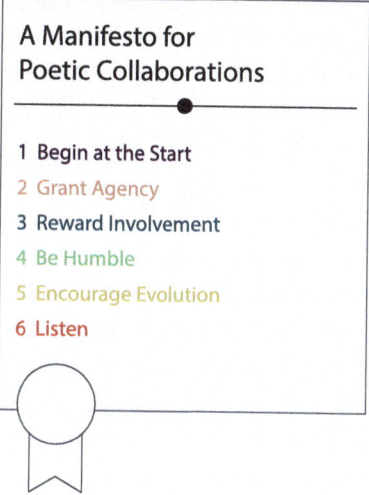

to one another. Do not assume what people want, ask them what they need. Create
space and opportunities for listening to the needs of your collaborators and make it
clear that their voices are both valued and necessary.

Exercise 7.1: apply this manifesto

Take an interdisciplinary collaboration that you have been involved in, or of
which you are aware (for example one of those mentioned in Sect. 7.1) and
apply the manifesto shown in Fig. 7.1. Did this collaboration begin at the start?
Was agency granted to all collaborators? How was involvement rewarded?
Were the lead organisers / instigators humble in their approach? Was the collaboration encouraged to evolve, and to what extent did the people involved listen
to one another?

   If the collaboration did not adhere to all six points of the manifesto, where
did it fall down and why? How might the collaboration have better incorporated
these action points, and would it have been improved as a result?

7.3   Case Study 1: Experimental Words

Experimental Words [17] is an ongoing collaboration that pairs together scientists
and poets to explore the liminal spaces between science and poetry. This project was
originally conceived and developed by myself and my long-term collaborator, the
award-winning UK poet Dan Simpson [18]. It began life at the 2015 Manchester

Science Festival, where we received a small amount of funding to pair together five scientists and five poets to create interdisciplinary performances for a one-off live event. The project expanded into three other UK cities (Edinburgh, London, and Canterbury) as a result of funding from both Arts Council England and the National Environmental Research Council, and a US edition also took place in Chicago as part of the Chicago Botanic Garden's 2018 'Unearth Science Festival'.

In 2021, further funding from Arts Council England was secured, enabling the creation of a spoken word album, featuring the work of ten scientist and poet pairings. The album (which is available via steaming services and as a free digital download) is also accompanied by three short videos created by UK filmmakers, and an algorithmically programmed and fully interactive album cover.

The initial aim of the Experimental Words project was to create interdisciplinary explorations of the work and research of both sets of practitioners, but to what extent has it successfully adhered to the six actions points of the manifesto described in Fig. 7.1?

7.3.1 Begin at the Start

Experimental Words has always existed as a fully collaborative project between Dan and me. Both of us have been involved from the very outset in the development of the collaboration, and have contributed equally to the grant writing, project management, and delivery. Initially when the project began, Dan was the 'poet' and I was the 'scientist', but these roles have since become much more fluid. For example, while we still retain our own individual networks for enlisting scientists and poets to work with us on this project, it is no longer the case that Dan only recruits the poets and I only recruit the scientists.

In terms of the collaborations that take place between the participating scientists and poets, Dan and I have always ensured that the poets are not 'simply' writing poems about the scientists' research, but rather that the paired collaborators are using both science and poetry to explore each other's work through an interdisciplinary lens. In some instances this has resulted in creating poetry that is influenced by published scientific research, but in many other cases the collaborations have resulted in new understandings and conceptualisations. This has been made possible because the pairings were introduced to one another at the start of the project rather than at the end.

Where Experimental Words could perhaps continue to improve, with regards to this action point of the manifesto, is by involving our collaborators in some of the earlier decision-making processes. This could include helping to generate future research funding or legacy development, although doing so would need to present clear benefits for all those involved.

7.3.2 *Grant Agency*

As discussed in Sect. 7.3.1, the roles that Dan and I have occupied in this project have been very fluid, and in terms of my own personal development as a poet and interdisciplinary researcher this has been invaluable. I know that the collaboration has also afforded Dan the opportunity to further develop his own practice, which in turn has led to more opportunities for him to work with other scientists in developing additional interdisciplinary projects.

With regards to the participating scientists and poets, we never prescribed the ways in which people could be involved, or the topics that should be covered. All that we asked was that both voices were present in the finished piece (i.e. either the live performance or the spoken word track). In some instances this involved performances from both contributors, whereas others used only one voice for the performance itself, and some incorporated multiple collaborators that extended beyond the initial pairings that Dan and I had facilitated. Several of the participants that were involved with Experimental Words would self-identify as being both a scientist and a poet, and they were free to contribute to the project in a manner that they found to be the most beneficial for them.

Furthermore, as a result of Experimental Words, several of the participants went on to re-define their own practice and research beyond the constraints of a single discipline, as evidenced from the following feedback gathered from one of the participating scientists:

> It gave me an understanding of how to create poetry, how to bridge the gaps between disciplines, gave me a creative outlet when my work environment and workload were not so great and the joy in my subject had been utterly lost.

In granting agency to the participants however, there were occasions where individuals required more guidance than was being provided. In developing this project Dan and I tried to create a facilitated process in which we were not overly prescriptive with what the outcomes might involve, but this was not always appreciated, as observed by one of the scientists who we worked with:

> I'm not sure if a few more facilitative meetings may also have helped or just kept up the information flow

In the same set of pairings, one of the poets made this observation:

> Halfway through I would have said more guidance, but I now know why you didn't – the range of poetry that was created was incredible.

These comments would seem to indicate that alongside greater agency, the participants could also be better informed (and involved—see Sect. 7.3.1) with regards to the process, to ensure that they had the most positive and beneficial experience possible.

7.3.3   Reward Involvement

Wherever possible we have tried to find funding to pay the poets for their involvement in Experimental Words, although for one of the earlier versions of the project this was not possible. In this instance we were able to offer some of the poets paid roles in future collaborations, while for others we appeared as (unpaid) collaborators on their own projects and/or used Experimental Words to signal share upcoming publications, gigs, and other events with which they were involved.

We did not reward the participating scientists with any financial incentives, and in almost all cases we were able to work with researchers who were in fulltime, paid employment. Despite this lack of financial reward, it is clear that the participating scientists benefitted greatly from the collaboration, as is evidenced from the following feedback:

> It was completely awesome. Lifechanging. I became a poet and kicked off a really good friendship. My life is richer

> I feel very grateful to have been invited to take part in Experimental Words. It was not only a chance to be involved in a new (to me) creative activity, but also offered an opportunity to reflect on my research as it is interpreted by a poet, which was invaluable.

> I learned so much, it was a lot of fun, and I hope to have contributed to science communication in a way which I would otherwise not have been able to do.

This approach was also somewhat justified by the following piece of feedback that we received from one of the participating scientists, when they were asked if anything could have been done to improve their experience:

> I do not know whether or not the poets were paid, but they should have been. Academics in permanent contracts (which is not all of us, but which applies to me) receive a regular salary, irrespective of their involvement in such projects. Poets (not all, but many) are effectively freelancers.

For future versions of Experimental Words, we will aim to state the potential benefits more explicitly to all of our collaborators, grounded in the feedback that we have received from previous participants. We should also find more specific ways to reward (and thus include) those participating scientists that are not on permanent contracts, as limiting ourselves in such a manner is denying an opportunity for greater diversity.

7.3.4   Be Humble

At every stage of the Experimental Words process we have solicited feedback from our collaborators to see how we can improve the project. Presented below are three pieces of feedback that were received following the creation of the spoken word album, and the ways in which we might use this to improve our approach:

> The deadlines were a little tight so that added to the stress at times.

Based on this piece of feedback, and in line with the recommendations suggested in Sect. 7.3.1, future versions of this project will now involve sharing the project timeline with collaborators at a much earlier stage than we have done previously, while also allowing for greater flexibility and contingency with deadlines.

Perhaps the organisation could have been a bit more ... er... organised

This feedback suggests that we should be much clearer in our approach. While several other participants praised our organisational skills, this was clearly not unanimous, and as such we need to make sure that we are meeting the individual needs of all the participants; for example by offering more one-to-one facilitation sessions where required. On a personal note, I found this piece of feedback to be particularly humbling, as I normally pride myself on my organisational skills. However, it is evident that I need to improve the way in which these skills are communicated and to be sensitive to the fact that maybe I am not as organised as I like to think.

It would have helped if a course was organised for the scientists on how to write poem or incorporate their work into poetry.

While this suggestion is not necessarily aligned to the original aim of the collaboration, it should not be ignored. On its own merit, this is a very good suggestion and one that presents an excellent opportunity for developing the legacy of the project, as will be discussed in Sect. 7.3.5.

7.3.5   *Encourage Evolution*

In addition to the evolution that has been made possible by being humble and listening to where we could make improvements (see Sect. 7.3.4), the Experimental Words project has undergone several stages of evolution, based on feedback and evaluation. One of the biggest forms of evolution is the way in which the project has been framed. Initially Experimental Words was called 'Science Slam', and while the format involved collaborative participation from the outset, this naming convention was later deemed to be inappropriate, especially with regards to the expectations of a viewing audience. As discussed in Chap. 3, a poetry slam is normally run as a competitive event, which is incongruent with the ethos of the paired collaborations that we helped to facilitate. While there is a history of science slams as a form of science communication [19], and as a means of developing the science communication skills of researchers [20], naming our collaborative project as such gave the impression that we were facilitating some kind of 'talent show', which is not the case. The Experimental Words branding evolved as a result of engaging with our collaborators to select a name that was more representative of the project.

The decision to make a spoken word album was also an important evolutionary step for this project, arising from the realisation that we had done a poor job of capturing the previous live performances. Reflecting on this need to document the process helped us to develop the concept of the album (and supporting videos), and to successfully apply for the funding that made this possible.

As discussed in Sect. 7.3.4, some of the feedback from one of our collaborators has raised the potential to evolve Experimental Words beyond an overreliance on Dan and myself. One of the limiting factors of this project is the need for us to initially facilitate the pairings of scientists and poets. However, by creating a suite of tools and resources for others to use, as well as potentially a networking hub for potential collaborators to meet, we could develop Experimental Words to be less reliant on the two of us, which in turn would open up many more opportunities for the project to evolve.

7.3.6 Listen

As demonstrated in Sect. 7.3.4, throughout Experimental Words we have created space for collaborators to share their voices; listening, responding, and where appropriate making actionable change. There are many things that we could still improve, and as the project evolves, we try to learn from our mistakes and to create a more equitable, fair, and mutually beneficial collaboration. The aspect of this project that I think we have succeeded with to the greatest extent is the sense of co-ownership that we have fostered. Dan and I have helped to facilitate Experimental Words, but as is evident from this final piece of feedback from one of the participating poets, it is a project that belongs to all of the collaborators, and one to which a sense of pride and achievement is firmly attached:

> Thank you for the opportunity and for thinking of me to be a part of this project. [...] we have created something very special.

Exercise 7.2: listen to Experimental Words

Listen to the spoken word tracks on the Experimental Words album [17] and read the accompanying sleeve notes that detail the experiences of each pairing during their collaborations. To what extent do you think this project is aligned with the manifesto presented in Fig. 7.1, and how do the experiences of the collaborators support this?

7.4 Case Study 2: Consilience

Consilience [21] is the world's first peer-reviewed science and poetry journal. It is a quarterly, online publication that publishes poems and artwork about science all linked by a common theme (which changes with each issue), submitted by creatives from all backgrounds. *Consilience* was set up to create a space for the people working on the overlaps between the sciences and the arts, and it is free to both read and submit work to.

In the majority of poetry journals, a poem is either 'acceptable' or 'unacceptable' at the stage of submission. *Consilience* instead takes inspiration from the peer review system of scientific journals [22] and uses this to provide the support that creatives need to develop both their craft and their identities. Only those poems that fall outside the scope of *Consilience*, the theme of the individual issue, or which are at odds with the inclusivity statement of the journal, are rejected out of hand. Every other submission is sent to two independent reviewers, who provide feedback on the poem. An editor then synthesises these reviews and collaborates with the poet to develop the piece for publication in the journal. This might involve re-thinking some elements, or it might be as simple as saying 'Keep everything exactly as it is!' However, every step of this process is a dialogue, and nothing is published without the explicit consent of the poet. An overview of this process is shown in Fig. 7.2.

Consilience is run by a team of over 70 volunteers from across the world and is currently read by about 5,000 people per month. I initially set up *Consilience* because I realised that a lot of my work in the field of science communication through poetry was limited by me as an individual, and that by creating a platform to help develop the work of others in this field I could hopefully break free from my own limitations. With that in mind, I will now discuss the development of *Consilience* according to the manifesto presented in Fig. 7.1.

Fig. 7.2 An overview of the Consilience process, from submission to publication

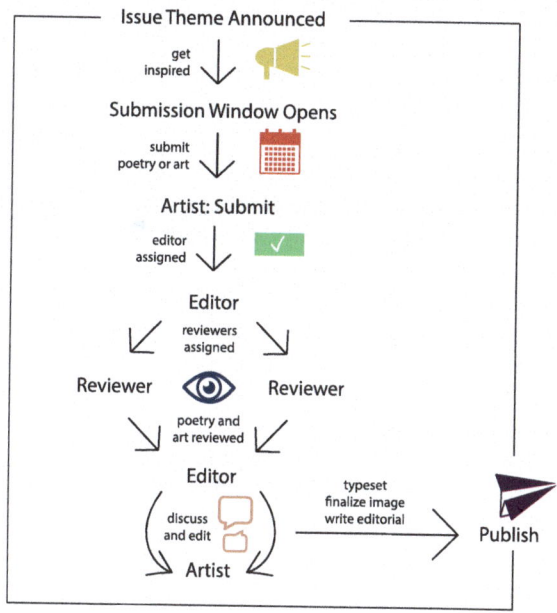

7.4.1 Begin at the Start

After having the initial idea for the *Consilience* journal, I contacted several colleagues who also operate in the spaces between science and poetry (a mixture of scientists, science communicators, poets, and other creatives) and asked them if they would like to be involved in the project, and to help define what it might achieve. From these early collaborative meetings we co-created the purpose of the journal, provided a framework for engagement, and determined the logistics for how work would be submitted, reviewed, and prepared for publication. It was decided very early on that the business communication platform Slack [23] would be the most efficient way for us to collaborate across multiple time zones, with Google Workspace [24] used to help co-create the various documents, spreadsheets, and forms that were necessary to facilitate the launch and subsequent upkeep of the journal.

Consilience is a good example of an interdisciplinary collaboration between scientists, poets, and other creatives, where the collaboration began at the very start of the project, and through which multiple voices were both present and platformed. Perhaps the biggest limitation for open, interdisciplinary collaboration at the start of *Consilience* was the fact that I initially reached out to scientists and poets in my own networks. While this has since been rectified (as my collaborators have gone on to invite their own, more varied networks to participate) this approach limited the diversity of those involved at the very beginning.

7.4.2 Grant Agency

During the initial conception of *Consilience* the main roles that I thought would be needed were those of reviewers and editors. However, from the very beginning several other people also wanted to contribute to the collaboration in alternative ways. As such we were joined by designers, web assistants, and social media experts all of whom have since helped to develop *Consilience* in new and exciting directions.

One of the biggest ways in which this collaboration has afforded agency to those involved is in the development of ConciliARTe (from the Spanish 'conciliar', meaning to reconcile or agree on something, and 'arte', being art) as an inclusive space for people's explorations between audio-visual art and science to sit alongside the poetry that is published in the journal. By granting agency to those collaborators who wanted to contribute in a way that was not specifically related to poetry, *Consilience* has been able to reach new audiences and to further consider how poetry and other artforms can work together to effectively communicate science.

Similarly, while I had initially thought that a Twitter profile might be useful for the journal, several of my collaborators decided that we should also create an Instagram account and a Facebook Group, to help diversify the reach of the journal and to create a space in which the readers and contributors could interact and engage with one another.

Finally, all members of the *Consilience* team are encouraged to engage with the journal in a way that aligns with their other interests and commitments. This means that if a reviewer is unable to contribute for a couple of issues, they are still made to feel part of the team. Similarly, those people who want to experience (or develop) new roles within the team are encouraged to do so and are supported in their development through peer-to-peer learning.

7.4.3 Reward Involvement

The biggest failing of *Consilience* to date is that it is currently unable to pay the reviewers, editors, and other collaborators for their time and contributions. This is largely because we have yet to develop a self-sustaining model that can generate some form of income while also keeping the magazine free to both read and publish in.

Besides the lack of financial incentive, there are however several ways in which the members of the *Consilience* team are rewarded. These include peer-to-peer learning, networking opportunities, the platforming of their other work and activities, and the opportunity to develop nascent skills in fields that are tangential to their usual work or employment.

In joining *Consilience* all members are presented with a 'Welcome Document' that outlines the different roles in the team, as well as the rough time commitments that are attached to each of these roles. As discussed in Sect. 7.4.2, all collaborators can vary the extent of their involvement, and from the various editorial meetings and the discussions that I have with each of the team members it is apparent that everyone who is involved in *Consilience* feels as though the benefits match their efforts. This is evident both from the way in which the *Consilience* team highlight their involvement in the journal through their CVs and social media profiles, and also from the unsolicited feedback that I regularly receive, a selection of which is shown below:

I'm just really honoured to be a part of the team.

It's a lot of pleasure for tiny effort on my end

We have many brilliant things that can come out of this, most importantly a platform and an ecosystem for poets to create and present great poetry that will help communicate science better.

In addition to the *Consilience* team, it is clear that the poets who submit their work to the journal feel suitably rewarded, not only in having their work published, but also through the engagement with their work that is afforded by the peer review process. The following two pieces of feedback are representative of many similar pieces of correspondence I receive from our published poets:

The detailed feedback from the *Consilience* reviewers has been invaluable in refining this poem and indeed sharpening the scientific message behind it, which is also important from an educational perspective.

The close reading I received from the editors, and the entire editorial process, has been rewarding and enjoyable.

7.4.4 Be Humble

Part of the continued success of *Consilience* has been driven by encouraging collaborators to experiment with their own ideas for the journal (see Sect. 7.4.5), and to not be limited by what I had originally perceived might be the journal's direction of travel. In particular, the development of ConciliARTe is not something that I had envisioned, and yet this is now an extremely successful and integral part of the journal which has resulted in many interdisciplinary collaborations that have extended beyond *Consilience* itself.

During the development of *Consilience* I have made many mistakes, but each time I have tried to learn from them rather than to become defensive or overprotective. For example, one of the first promotional videos that I created for the journal was rightly highlighted on social media as not having subtitles. As a result of this I taught myself how to embed subtitles into videos, which has led to the work of both *Consilience* and my other creative (and academic) output being more accessible. Similarly, some of the documents that are used by the *Consilience* team to facilitate the review process were highlighted as being somewhat inaccessible to people with lower levels of digital literacy. As such I collaborated with a number of individuals to improve the accessibility of these documents, including creating worked examples and FAQs that helped to both simplify and streamline the process. The result of this was that everyone benefited, even those with very high levels of digital literacy.

By fostering a collaborative environment in which respectful feedback is encouraged and acted upon, I have hopefully helped to move beyond the limitations of my own ideas. Inviting opinions, critiques, and comments on my own suggestions has helped others to suggest and develop new ways for the journal to grow. This would not have been possible if I had possessively implemented only my ideas for the journal, without any consultation with the other members of the *Consilience* team.

7.4.5 Encourage Evolution

One of the greatest strengths of *Consilience* has been in promoting a bottom-up culture in which all ideas are encouraged. Enabling the evolution of the journal in such a way has resulted in the development of many new opportunities. Aside from ConciliARTe, two of the most interesting have been the virtual exhibitions and the Science & Poetry Café.

The virtual exhibitions are a series of online installations in which poetry and artwork from the journal are paired together and curated under a single theme (e.g. 'Medicine' or 'The Anthropocene'). These exhibitions are free to attend and can be viewed via the *Consilience* website. Each exhibition is also accompanied by a launch, a free online event in which the selected poets and artists talk about their work and add additional context to the pairings. Feedback from both the exhibitors and the audience have been used to develop this format and to make it more inclusive. For example, while the launch event for the first virtual exhibition ('Medicine') was well attended and received very positive feedback, it became clear that we had not suitably briefed the exhibitors about their role in the event itself. This was rectified for future launch events and meant that people were able to prepare their contributions in advance, which greatly improved the quality and professionalism of the experience.

The Science & Poetry Café was developed as an online space for people from across the world to share their science-related poetry in an inclusive and nurturing space. For similar reasons to those discussed in Sect. 7.3.5, these events have more in common with an open mic poetry event (see Chap. 3) than a poetry slam. At these events (which are free to attend and are hosted using the Zoom video teleconferencing software service), more established poets are invited to perform a short set of science-related poetry, with several slots available for the other participants to read their own poems. These cafés have tended to attract large audiences (several hundred participants), and key to their success has been in providing a space for participants to discuss their experiences after the event, in a café-style format. As with the virtual exhibitions, we have used feedback from both the participating poets and the audience to evolve the experience. For example, after the first café, it became clear that the 'live transcriptions' were not particularly accurate, and so instead poets at all future events were encouraged to 'screen share' their poems as they read them out loud, so that the audience could better follow their performance. Similarly, following feedback from the first Science & Poetry Café, 'menus' were produced for all attendees, which contained the contact details of the poets (e.g. websites, social media handles) so that people could more easily get in contact with them after the event.

Several other ideas have been developed since the inception of *Consilience*, some of which have been a great success and others of which have not materialised as initially expected. However, promoting a culture in which all collaborators are encouraged to develop the journal has helped to re-define both the role and purpose of *Consilience*, while supporting those who want to further explore and expand science communication through poetry.

7.4.6 Listen

Throughout the development and continued existence of *Consilience* we have ensured that there are various spaces in which we can all listen to each other's needs and experiences. While there is a great diversity within the team in terms of both personalities and discipline expertise, there is a mutual respect that helps to platform all voices. On a personal level I try to accommodate the needs of all of the *Consilience* team, which is a learning experience that I continue to benefit from. As discussed in Sect. 7.4.3, the greatest failing of this collaboration to date has been in generating income to pay for the contributions of the collaborators. However, I am convinced that as an interdisciplinary team this is a problem that we will be able to solve, and one which is no longer restrained by my own limitations.

Exercise 7.3: read Consilience
Read the latest issue of *Consilience* [21] and have a look at the way in which the journal is structured. To what extent do you think this project is aligned with the manifesto presented in Fig. 7.1, and how does the journal help (or hinder) interdisciplinary collaborations more generally.

7.5 Summary

In this Chapter, I have developed a six-point manifesto for interdisciplinary collaborations between scientists and poets. In doing so I have demonstrated why such a manifesto is necessary, and why it takes the form that is presented in Fig. 7.1. I have also applied this manifesto to two well-established interdisciplinary collaborations to demonstrate how it works in practice, highlighting the merits and the areas for improvement in both Experimental Words and *Consilience*. Having read this Chapter you should be well equipped for developing your own interdisciplinary collaborations between poets and scientists that are fair, equitable, and beneficial for all participants. In the next and final chapter I will summarise the learnings from this book and also discuss how you might further use some of the ideas that are presented here to develop your own science communication through poetry.

7.6 Suggested Reading

Cross-Pollinations: The Marriage of Science and Poetry by Gary Paul Nabhan [25] presents a series of essays that explore how interdisciplinary collaborations between science and poetry (and also the arts more generally) can lead to new understanding; for example how the structure of a specific poem inspired the author to develop a fundamental new research model in the field of agricultural biology. Similarly, 'Poetry as a creative practice to enhance engagement and learning in conservation science' [26] and 'Creative expression of science through poetry and other media can enrich medical and science education' [27] present detailed accounts of how poetry can help to develop fundamental research through meaningful collaborations across the disciplines. Finally, *Artscience: Creativity in the post-Google Generation* by David Edwards [28] explores how crossover learning between disciplines helps to provide a catalyst for innovation, including how ideas can be effectively translated within and outside of academia.

7.7 Further Study

The further study section in this Chapter is designed to help you develop your own interdisciplinary collaborations between poets and scientists, and to consider the role that a manifesto can play in helping these to achieve success.

1. **Join Consilience**. Visit the *Consilience* website [21] and consider submitting a poem to one of our forthcoming issues. This might be one of the poems that you have developed in the other chapters in this book. You might also consider connecting with us across social media, and maybe even joining the team as a reviewer, editor, or another role that you would like to develop. All voices are welcome, and we would love to hear from you.
2. **Create a collaboration**. Taking inspiration from the two case studies that were presented in this Chapter, can you develop an interdisciplinary collaboration between scientists and poets of your own? How might you ensure that this collaboration is aligned with the manifesto presented in Fig. 7.1, and what will be your role in its development, delivery, and evolution?
3. **Create a manifesto**. Is the manifesto that I have presented in this chapter suitable for your own needs? Have I missed something out or is there something which you do not think is necessary for effective interdisciplinary collaborations? Using the manifesto of Fig. 7.1 as a starting point adapt it as necessary, being sure to justify each change and to also present clear examples so that others might follow your manifesto as and when required.

References

1. Cosgrave EJ, Kelman I (2017) Performing arts for disaster risk reduction including climate change adaptation. In: Kelman I, Mercer J, Gaillard JC (eds) The Routledge handbook of disaster risk reduction including climate change adaptation. Routledge, Abingdon
2. Cressey D (2013) Arts: framing change. Nature 497:187. https://doi.org/10.1038/497187a
3. Triscott N (ed) (2014) The arts catalyst reader volume 1. The Arts Catalyst, Sheffield
4. Buckland D (2012) Climate is culture. Nat Clim Change 2:137–140. https://doi.org/10.1038/nclimate1420
5. Blakinger JR (2016) The aesthetics of collaboration: complicity and conversion at MIT's center for advanced visual studies. Tate Papers 25 (Spring)
6. Moura JM, Llobet J, Martins M et al (2018) Creative approaches on interactive visualization and characterization at the nanoscale. In: Brooks A, Brooks E, Sylla C (ed) Interactivity,

Game creation, design, learning, and innovation. ArtsIT 2018, DLI 2018. Lecture Notes of the Institute for Computer Sciences, Social Informatics and Telecommunications Engineering, vol 265. Springer, Cham

7. Paterson SK, Le Tissier M, Whyte H et al (2020) Examining the potential of art-science collaborations in the anthropocene: a case study of catching a wave. Front Mar Sci 7:340. https://doi.org/10.3389/fmars.2020.00340

8. Hoover KM, Lee J, Hamrick T (2020) Community engagement in science through art (CESTA) summer program. J Chem Educ 97(8):2153–2159. https://doi.org/10.1021/acs.jchemed.9b01101

9. Rosin M, Wong J, O'Connell K et al (2021) Guerilla science: mixing science with art, music and play in unusual settings. Leonardo 54(2):191–195. https://doi.org/10.1162/leon_a_01793

10. Koek A (2017) In/visible: the inside story of the making of Arts at CERN. Interdisc Sci Rev 42(4):345–358. https://doi.org/10.1080/03080188.2017.1381225

11. Forget B (2021) Women with impact: taking one small step into the universe. Leonardo 54(1):63–70. https://doi.org/10.1162/leon_a_01985

12. Guthridge GG (2021) Beyond STEAM: integrative inquiry in the Antarctic. J Geosci Educ 69(2):100–105. https://doi.org/10.1080/10899995.2020.1735012

13. Kristiansen DM (1968) What is Dada? Educ Theatr J 20(3):457–462. https://doi.org/10.2307/3205188

14. Hate Socialist Collective (2009) Leave the manifesto alone: a manifesto. Poetry 193(5):452–454

15. Einstein A, Russel B (1955) The Russel-Einstein manifesto. London

16. Open Science Repository (2012) Scientific Manifesto Based on the Philosophy of Science of Karl Popper. Open Sci Repos Philos 1(09). https://doi.org/10.7392/journal.philosophy.0000071

17. Illingworth S, Simpson D (2021) Experimental words. https://experimentalwords.bandcamp.com/. Accessed 10 Dec 2021

18. Simpson D (2021) Dan Simpson poet. https://www.dansimpsonpoet.co.uk. Accessed 10 Dec 2021

19. Niemann P, Bittner L, Schrögel P et al (2020) Science slams as edutainment: a reception study. Media Commun 8(1):177–190. https://doi.org/10.17645/mac.v8i1.2459

20. Zarkadakis G (2010) FameLab: a talent competition for young scientists. Sci Commun 32(2):281–287. https://doi.org/10.1177/1075547010368554

21. Illingworth S (2021) Consilience. https://www.consilience-journal.com. Accessed 10 Dec 2021

22. Alberts B, Hanson B, Kelner KL (2008) Reviewing peer review. Science 321(5885):15. https://doi.org/10.1126/science.1162115

23. Perkel JM (2017) How scientists use Slack. Nature 541:123–124. https://doi.org/10.1038/541123a

24. Akcil U, Uzunboylu H, Kinik E (2021) Integration of technology to learning-teaching processes and google workspace tools: a literature review. Sustainability 13(9):5018. https://doi.org/10.3390/su13095018

25. Nabhan GP (2004) Cross-pollinations: the marriage of science and poetry. Milkweed Editions, Minneapolis

26. Januchowski-Hartley SR, Sopinka N, Merkle BG et al (2018) Poetry as a creative practice to enhance engagement and learning in conservation science. Bioscience 68(11):905–911. https://doi.org/10.1093/biosci/biy105

27. Brown SA (2015) Creative expression of science through poetry and other media can enrich medical and science education. Front Neurol 6:3. https://doi.org/10.3389/fneur.2015.00003

28. Edwards DA (2008) Artscience: creativity in the post-google generation. Harvard University Press, Cambridge

Chapter 8
Conclusions

8.1 Key Concepts

In this final chapter, I will summarise the key conclusions from this book, and then make some recommendations for what you might decide to do next in your own journey of communicating science through poetry.

There are four key concepts that I have returned to throughout this book, which I will now summarise in turn (Fig. 8.1).

Poetry, like science, should be for everyone

Science can be an incredibly exclusive and exclusionary 'club', one that suffers from a lack of inclusion [1], equality [2], and retrospection [3]. Addressing this lack of diversity should be paramount to anyone who is involved in either science or science communication, not only because it is ethically the 'right thing' to do, but because ultimately greater diversity results in better science [4]. Poetry possesses the potential to help diversify both science and scientific discourse, but to do so, it also needs to address its own issues of inclusivity. Central to this is the concept that many people think that poetry is 'not for them', a notion that is at times reinforced by the dominance of a certain 'type' of poetry in western education and culture. Working with publics and individuals to showcase and create poetry that they enjoy, which relates to them, and which extends beyond Eurocentric notions of aesthetics helps to address this. The more people we can empower to read, analyse, and write poetry, the more potential there is to turn this lens onto science, thereby increasing the potential for diverse solutions to global, interdisciplinary problems such as disease transmission, water scarcity, and the climate crisis.

Poetry is a powerful way of communicating science

Science communication exists on a spectrum: from dissemination to dialogue. While dialogue is likely to be the most effective way of helping to truly diversify science, there is still a need for science communication initiatives that exist across this spectrum. This book has discussed how poetry might be used to support a variety of

© The Author(s), under exclusive license to Springer Nature Switzerland AG 2022
S. Illingworth, *Science Communication Through Poetry*,
https://doi.org/10.1007/978-3-030-96829-8_8

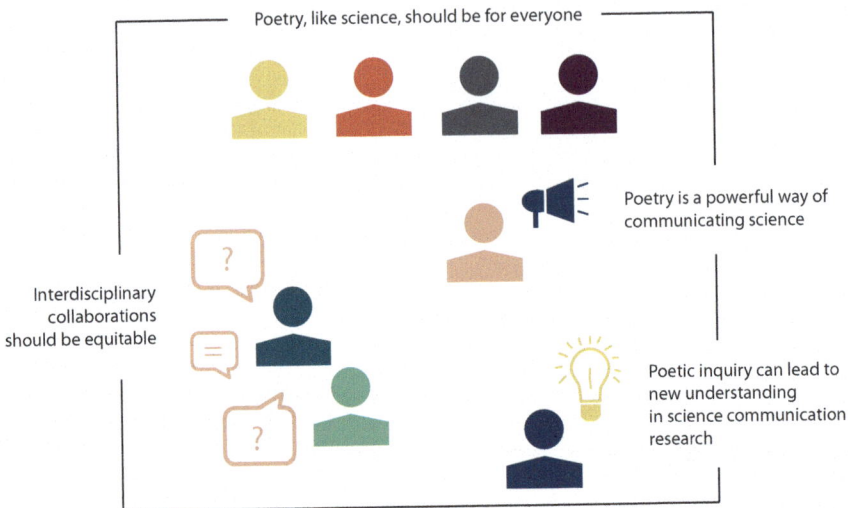

Fig. 8.1 The four key concepts of this book

such initiatives; from writing and sharing poems about a piece of scientific research (dissemination) through to running poetry workshops that help to bring together multiple publics to share their expertise and lived experiences in relation to a specific scientific topic (dialogue). Poetry affords scientific researchers and science communicators the opportunity to engage with these publics in a manner that is both inclusive and mutually beneficial, but only if we are prepared to listen to the needs of the publics with whom we are communicating.

Poetic inquiry can lead to new understanding in science communication research

Chapters 4 and 5 presented two distinct research methods for using poetic inquiry to interrogate new avenues in science communication research. These research methods are grounded in a scholastic body of work that is both rigorous and ethical, offering an insightful approach to formulating and answering original research questions. By detailing their provenance and providing several worked examples and case studies I have demonstrated how poetic inquiry might be used to continue to advance the still relatively nascent field of scholarly research in science communication.

Interdisciplinary collaborations should be equitable

Whether you are conducting a poetry workshop between a scientific and a non-scientific public, or organising an initiative that pairs scientists and poets, you should ensure that all voices are given an opportunity to be both heard and acted upon. True collaboration is a multidirectional process that extends beyond one person (or public) asking another to do something for them. It involves trust, agency, and benefits that are apposite for all collaborators. Be sure also to fully consider your own role in the collaboration. As I have stated many times in this book, it is important to be humble, but you should also try to avoid false modesty and any potential tendencies towards imposter syndrome [5]. Facilitating such collaborations is a very demanding task, and one that requires a unique skillset that can be both developed and celebrated. In considering your role in such collaborations, think also about what you stand to gain, and how it might contribute towards your own development in the process.

8.2 What You Do Next

You've read this book, studied the four key concepts that I hoped to communicate, and maybe even made a start on some of the suggested readings that I provided in each of the chapters. So what next (Fig. 8.2)?

Become more familiar with poetry

Read more poetry. Analyse more poetry. Write more poetry. Sections 8.3 and 8.4 provide some suggestions for expanding your poetic library, but this is only the start. If you have read a poem recently then what is it that you enjoy about it, and why? How does it relate to science? Does it relate to science? Why not? Who can you discuss this poem with, and what do they think about it?

In Chap. 2 I encouraged you to actively target the incubation period by writing poetry whenever you found yourself at an impasse in your scientific work. This is a very effective way of helping to familiarise yourself with poetry, but you might also extend it to other issues that need resolving in your professional and/or personal life. Keeping a regular poetic diary might seem like a large commitment, but it is a great way to help familiarise yourself with poetry. Challenging yourself to summarise a key event that has happened in your day/week/month/year will also help you to develop reflective skills (see Chap. 6) that you can utilise elsewhere. Similarly, if you encourage yourself to experiment with different poetic forms for each entry then you will quickly become accustomed to the opportunities and limitations that each of these present.

Fig. 8.2 What you should
do next

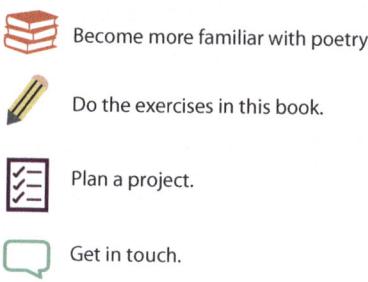

Become more familiar with poetry.

Do the exercises in this book.

Plan a project.

Get in touch.

Do the exercises in this book

I mean *really* do the exercises in this book. Go back to Chap. 1 and work your way through each of the exercises in turn. These exercises have specifically been designed to help familiarise you with poetry, and to then consider how you might use this knowledge to help communicate, interrogate, and diversify science. Even if you are only interested in part of this book, I still recommend that you work your way through each of the exercises in turn, as doing so will help to provide additional insight for what you plan to do next.

For example, if you are primarily interested in developing poetry workshops to initiate dialogue between scientific and non-scientific publics, then I would still recommend that you work your way through the exercises that are provided in Chaps. 4 and 5. Even though these chapters and exercises are primarily concerned with poetic inquiry as a form of qualitative science communication research, they will still provide you with valuable insight for how to develop more effective, equitable, and ethical poetry workshops. They might also give you some ideas for how (and why) you could develop these workshops into publishable research.

Similarly, if you are mainly interested in learning about how to implement poetic content analysis, then you should still complete the exercises in Chaps. 2 and 3, as doing so will help you to better develop an understanding of the potential relationships between science and poetry, which in turn will deepen your understanding of this research method.

Plan a project

After working your way through all of the exercises in this book, you should have an idea of the kind of science and poetry initiatives that interest you the most, and to which you are perhaps best equipped to develop, deliver, and evolve. Using these experiences as an inspiration, start to develop a new science and poetry project. This might be a chapbook of poems that explores ecological collapse, or a series of poetry workshops that aims to generate dialogue and action with regards to indigenous knowledge and earthquake predictions. Alternatively you might decide to use poetic transcription to give voice to the memoirs of a long-forgotten scientist or be keen to set up an interdisciplinary collaboration between poets and scientists to help formally recognise animals as sentient beings. Whatever it is that you decide to do, think carefully who your publics are for this project, and be sure to engage them as soon as possible.

Get in touch

If there is anything in this book that is unclear or which requires further comment, then please get in touch. Similarly, if you think that I have missed out something important, neglected an insightful reference, or even made an embarrassing grammatical or typographical error, then please get in touch. Finally, if you want to share your science poems, tell me how you are using poetry in your own research and practice, or discuss an opportunity for us to collaborate, then please get in touch. I love discussing science and poetry, and I am at my happiest when I am finding new ways to work with new people on new projects that broaden the limits of my own language and world. You can find me on Twitter @samillingworth, through my website www.sam illingworth.com, or via email at sam.illingworth@gmail.com. Please get in touch, and let's discuss how we can better communicate science through poetry, together.

8.3 Suggested Reading

Throughout this book I have suggested numerous books, articles, and reports that explore the intersections of science and poetry. This has also included a small number of poetry collections that are related to, or informed by, science. However, given that one of my strongest motivating factors for writing this book was to help people feel more comfortable with and inspired by poetry, I would like to dedicate this final set of suggested readings to poetry collections that have helped me to appreciate poetry. You might love the poems in these collections, or you might not care for them at all, but I strongly recommend that you give them a read, as they really help to redefine what is possible in this incredible medium.

Claudia Rankine's *Citizen: an American Lyric* [6] is a multi-award winning account of racial relations in the United States. Presented as a book-length poem, it offers a nuanced and troubling reflection on race, identity, and a sense of belonging. It is probably unlike any other poem that you will have read before, and it highlights how the medium is the perfect vehicle for giving voice to key societal issues that might otherwise go unspoken.

England: Poems from a School [7] is an edited collection from the Scottish poet Kate Clanchy. It presents poems that have been written by students from the Oxford Spires Academy, a small secondary school in the UK that has a large multinational cohort of students, who between them speak 30 different languages. The collection is both a wonderful celebration of youth and multiculturalism, and also an opportunity to hear about the heartbreak and suffering of migration from those whom it most affects.

Talking to the Dead [8] is a collection of poetry from the English poet and novelist Elaine Feinstein. The poems present a series of conversations with / reflections on her late husband and her relationship with him. It is an astonishing set of poetry where not a single syllable is wasted, and which showed me how even the briefest and most succinct of poems could convey a lifetime of emotion and thought.

Finally, *The Penguin Book of the Prose Poem* [9] edited by Jeremy Noel-Tod presents over 100 prose poems from across the last 150 years of the format. As well as an accessible introduction to the history and development of this form of poetry it features a wide-ranging selection of poets and their voices, from Charles Baudelaire in 1869, through to the latest works of Claudia Rankine and Sarah Howe. This anthology also contains what might be my favourite poem of all time 'The Colonel' by Carolyn Forché. Even now, as I read this poem for the hundredth time, I feel elated and indebted that somebody had written those words so that I might have the pleasure to read them.

8.4   Further Study

The further study section in this chapter is designed to help you reflect on the key learnings of this book, and to use them to develop your own science communication through poetry.

1. **Read some poetry.** Start off with the poetry collections that I have recommended in Sect. 8.3, and throughout the rest of the book. Some of these are anthologies that contain the work of several poets, so if you read a poem that you like, then check out their other work as well. Subscribe to some poetry journals (either online or in print) and follow a selection of poets on social media; Instagram is especially good for this. Remember that it is ok for you to dislike poetry that other people love, and for you to love poetry that others dislike. No one can tell you how you should feel about a poem—that is your right and privilege to discover for yourself.

2. **Share your poetry.** Having worked your way through the various exercises in this book you should now have written several science-themed poems. You may even have been inspired to write some additional poems of your own or to have created poetic transcriptions as a form of poetic inquiry. Think about who you can share these with to gather feedback, and how you might use this to develop both your poetry and also your voice as a writer. Remember that you don't have to make every change that is presented by each person who offers feedback but try to understand why they have made this recommendation in the first instance, and the rationale you might have for following or ignoring their advice.

3. **Find a collaborator.** Hopefully you now have lots of ideas for communicating science through poetry, via a range of different approaches. One of the best pieces of advice I can give is for you to find a collaborator to help make it a reality. Working with others helps to fill the gaps in your own skillset, introduces you to new audiences, presents new perspectives on your work, and creates opportunities for further collaborations. Such collaborators can also provide feedback on your writing, as discussed above. From my own experiences, such collaborations have also resulted in many lasting friendships, through which my life has become immeasurably richer as a result.

And finally, thank you for reading. If you have made it to the end of this book (via whatever path), then I am extremely grateful for your time, engagement, patience, and presence. I hope that as a result of your reading you feel encouraged and empowered to use poetry to help communicate and diversify science. Please do get in touch if you have any ideas, projects, or problems that you would like to work on together. Be safe, be well, and in the words of Maya Angelou remember: 'everything in the universe has a rhythm.'

References

1. Puritty C, Strickland LR, Alia A et al (2017) Without inclusion, diversity initiatives may not be enough. Science 357(6356):1101–1102. https://doi.org/10.1126/science.aai9054
2. Urry M (2015) Science and gender: scientists must work harder on equality. Nature 528(7583):471–473. https://doi.org/10.1038/528471a
3. Rodriguez AJ (2015) What about a dimension of engagement, equity, and diversity practices? A critique of the next generation science standards. J Res Sci Teach 52(7):1031–1051. https://doi.org/10.1002/tea.21232
4. Medin DL, Lee CD (2012) Diversity makes better science. APS Observer 25(5)
5. Mullangi S, Jagsi R (2019) Imposter syndrome: treat the cause, not the symptom. JAMA 322(5):403–404. https://doi.org/10.1001/jama.2019.9788
6. Rankine C (2014) Citizen: an American Lyric. Penguin Books, London
7. Clanchy K (ed) (2018) England: poems from a school. Picador, London
8. Feinstein E (2011) Talking to the dead. Carcanet, Manchester
9. Noel-Tod J (ed) (2018) The penguin book of the prose poem: from Baudelaire to Anne Carson. Penguin Books, London

Index

The manufacturer's authorised representative in the EU is Springer
Nature Customer Service Centre GmbH, Europaplatz 3, 69115 Heidelberg,
Germany. If you have any concerns regarding our products, please
contact ProductSafety@springernature.com

Printed and bound by CPI Group (UK) Ltd, Croydon, CR0 4YY

29/04/2026

02099459-0003